BREWING TECHNOLOGY

Edited by **Makoto Kanauchi**

Brewing Technology
http://dx.doi.org/10.5772/66269
Edited by Makoto Kanauchi

Contributors

Djoulde Darman Roger, Bayoï James Ronald, Jang Eun Lee, Jae-Ho Kim, Antonio Morata, María Jesús Callejo, María Carmen González, Sonia Ancuta Socaci, Anca Corina Farcas, Francisc Vasile Dulf, Maria Tofana, Elena Mudura, Liana Claudia Salanta, Dan Cristian Vodnar, Marilena Budroni, Giacomo Zara, Maurizio Ciani, Francesca Comitini, Nermin Gozukirmizi, Elif Karlik, Takanori Yoshikawa, Shin Taketa, Makoto Kanauchi, José Pedro Santos, Jesús Lozano, Manuel Aleixandre

Published by InTech
Janeza Trdine 9, 51000 Rijeka, Croatia

Publishing Process Manager Romina Rovan
Technical Editor SPi Global
Cover InTech Design team

First published July, 2017
Printed in Croatia
Legal deposit, Croatia: National and University Library in Zagreb

Additional hard copies can be obtained from orders@intechopen.com

Brewing Technology, Edited by Makoto Kanauchi
p. cm.
Print ISBN 978-953-51-3341-4
Online ISBN 978-953-51-3342-1

For EU product safety concerns:
IN TECH d.o.o., Prolaz Marije Krucifikse Kozulić 3, 51000 Rijeka, Croatia,
info@intechopen.com or visit our website at intechopen.com.

Meet the editor

Makoto Kanauchi is a professor in the Department of Food Management, Miyagi University. He has been in the Department of Food Management, Miyagi University, since April 2005. Earlier, he was employed at the Institute of Food Science in Fuji Oil Corporation in Moriya, Ibaraki, Japan, as a researcher (August 2003–March 2005). Dr. Kanauchi also worked in Prof. Charlie Bamforth's Laboratory in Food Science and Technology, University of California, Davis, California (November 1999–May 2003). In March 1999, he received his PhD degree in Bio-regulation from Tokyo University of Agriculture. Dr. Kanauchi received his MS degree in Brewing and Fermentation Science in March 1996 and graduated from Tokyo University of Agriculture in Tokyo, Japan.

Contents

Preface

Beer has a long history. Beer was brewed in ancient Mesopotamia in 3000 BC. In today's world, beer is an extremely popular alcoholic beverage consumed worldwide. Its annual global consumption was 146 million kL in 2015. People in China, who consumed 43 million kL of beer in 2015, collectively consumed more beer than people in any other country. The second leading country for consumption is the US, with 22 million kL of beer consumed, followed by Brazil, Germany, and Mexico. These five countries collectively consume more than 60% of all the beer produced worldwide. Brewing technology has occupied an extremely important position in food industry. Moreover, because beer and other alcoholic beverages have been subject to taxation in many countries, they account for a large percentage of all tax revenues. Huge brewery companies truly have great social importance.

This book includes a collection of chapters widely related to the brewing industry.

The book is divided into the following five sections:

- Brewing materials
- Microbiology of brewing
- Brewing technology
- Application of brewing industry waste
- Quality control of beer

In the first section, describing brewing materials, the authors discuss two issues related to barley: factors responsible for converting barley plants into superior cereals occurring through gene transfer, gene editing, and molecular breeding and molecular mechanisms underlying leaf development in a medial-lateral direction. Moreover, oxidants have the potential of degrading beer. Oxidant enzymes such as thiol oxidase, ascorbate peroxidase, and ascorbic acid oxidase in malt are also discussed. Subsequent sections present discussion of brewing-related microbiology. The potential use of non-Saccharomyces strains in brewing processes is explained by two contributors. One describes *S. cerevisiae* strains isolated from different food matrices (i.e., bread, wine) and non-Saccharomyces starter strains in beer production, which are discussed for their potential uses in brewing. Another presents an interpretation of some non-Saccharomyces yeasts such as *Schizosaccharomyces pombe, Torulaspora delbrueckii, Lachancea thermotolerans, Saccharomycodes ludwigii,* and *Brettanomyces bruxellensis* for their use in brewing. The traditional local alcoholic beverage, nuruk, is introduced in the next section to highlight brewing technology in Korea.

In the section explaining brewing waste application, the authors discuss the use of brewing industry wastes to produce functional ingredients and various products with added value.

Finally, to discuss beer quality control, electronic noses are described as devices using sensors that measure volatile chemical compounds.

All chapters were written by professionals working in the brewing industry, at research institutes, and at universities. The issues all outline extremely interesting and important leading-edge brewing technology. This book is therefore appropriate for use by university researchers and brewing practitioners: brewers at breweries.

I would like to acknowledge all authors for kindly contributing their chapters. We are grateful to the Publishing Process Manager, Ms. Romina Rovan, for her assistance and to the information technology department for providing the requisite framework, which greatly enhanced the compilation of the book chapters.

Makoto Kanauchi PhD
Professor
Department of Food Management
Miyagi University, Sendai City, Japan

Section 1

Brewing Materials

Chapter 1

Narrow Leaf Mutants in the Grass Family

Takanori Yoshikawa and Shin Taketa

Additional information is available at the end of the chapter

http://dx.doi.org/10.5772/intechopen.68794

Abstract

Leaf morphology is critical for the survival of plant species. After a leaf primordium is initiated at the flank of shoot apical meristem (SAM), the development along the medial-lateral direction enlarges the leaf-blades, leading to the increase of photosynthetic activities. Thus, the revelation of mechanisms that control development across a leaf is quite important for plant breeding. A variety of narrow leaf mutants have been identified in the grass family, which includes particularly important crops in the world. Here, the molecular mechanisms underlying the leaf development in the medial-lateral direction are discussed as we introduce the three major groups of narrow leaf mutants in the grass family: (1) auxin-related mutants, (2) cellulose synthase-like D (CSLD)-related mutants, and (3) polarity-related mutants. The results obtained from these analyses could be directly applied to the breeding of major cereal crops such as maize, rice, and barley; therefore, they could contribute to the increase of food production.

Keywords: barley, rice, maize, leaf morphogenesis, mutant, gene expression

1. Introduction

Leaves are the major photosynthetic organs in plants. The light-capture efficiency significantly differs depending on the leaf shapes, angles, and arrangements in the canopy. Steeper leaf angle allows more light to penetrate to the lower leaves, leading to the increase of carbon gain through assimilation [1]. To avoid self-shading, leaf arrangement (phyllotaxis) is highly regulated by the plant hormone auxin [2, 3]. Since carbohydrates used in living activities are largely derived from the photosynthesis in plants, leaf morphology is critical for the survival of plant species.

A leaf primordium is initiated at the flank of shoot apical meristem (SAM), in which cells are maintained an indeterminate state by *class I knotted1-like homeobox* (*KNOX*) genes. The *Arabidopsis thaliana* genome includes four *class I KNOX* genes; *shoot meristemless* (*STM*), *brevipedicellus*

(*BP*), *KN1*-like in *Arabidopsis thaliana2* (*KNAT2*), and *KNAT6* [4]. *STM* is expressed throughout the SAM and induces cytokinin biosynthesis via *isopentenyl transferase7* (*IPT7*) activation and negatively regulates gibberellin biosynthesis via *GA 20-oxidase1* (*GA20ox1*) repression [5]. The resulting high cytokinin and low gibberellin ratio promotes meristem maintenance [6]. Such *STM* expression is downregulated by plant hormone auxin [2, 3]. Auxin is unique in its polar transportation mediated by influx carriers represented by AUXIN1 (AUX1) and LIKE-AUX1 (LAX) proteins, and efflux carriers represented by PIN-FORMED (PIN) and ATP-binding cassette B (ABCB) proteins [7]. Once transported to SAM, auxin flows to the peripheral young leaf primordia, creating an auxin maximum in the region where leaf primordia do not exist in the meristem. Such auxin localization downregulates *STM* expression, leading to the low cytokinin and high gibberellin ratio, which promote the switch from an indeterminate to a determinate state [8]. The loss-of-function of *PIN1* results in the malformed leaf development such as fused or cup-shaped leaves, suggesting that localized auxin accumulation in the meristem determines the radial position of leaf initiation [9].

In SAMs, STM also downregulates the expression of the MYB transcription factor *asymmetric leaves1* (*AS1*) and lateral organ boundaries domain (LBD) transcription factor *AS2*. When STM is repressed due to the auxin localization, AS1 and AS2, released from the negative regulation of STM, act together as a heterodimer to repress the expression of *BP*, *KNAT2*, and *KNAT6* to prevent cell fate from returning to meristem [10–12]. The loss-of-function of *AS1* resulted in the malformation of leaves due to the ectopic *BP* expression, which was enhanced with the additional loss-of-function of *auxin resistant1 (AXR1)* encoding a subunit of the related to ubiquitin1 (RUB1) activating enzyme that affects auxin responses [13]. These results suggest that the expression of *AS1* together with auxin localization plays a pivotal role in conferring leaf fate and promoting leaf development. Interestingly, slight *KNOX* expression remains in leaf primordia in species with compound leaves [14]. In tomato, the *class I KNOX* genes *tomato knotted1* (*TKN1*) and *TKN2* are expressed in young leaf primordia [15, 16]. The repression of TKN activity quickens the transition of the leaf primordia from the initiation to the secondary morphogenesis, suggesting that KNOX proteins are involved in the delay of leaf maturation and enable leaflet formation within leaf primordia [16].

The morphogenesis of sophisticated leaf organs with high reproducibility is achieved through the development in accordance with three axes; the proximal-distal, adaxial-abaxial, and medial-lateral directions (**Figure 1A–E**) [8, 17]. The development along the medial-lateral direction enlarges the leaf-blades, leading to the increase of photosynthetic activities. Thus, the revelation of developmental mechanism along the medial-lateral direction is quite important for plant breeding. So far, a variety of narrow leaf mutants have been identified in the grass family, which includes particularly important crops in the world. The results obtained from these analyses could be directly applied to the breeding of major crops such as maize, rice, and barley; therefore, they could contribute to the increase of food production. In fact, erect and narrow-leafed rice mutants led to the higher photosynthetic CO_2 uptake and improved yield in dense planting [18]. Recently, it was revealed that the Quantitative Trait Locus (QTL) controlling flag leaf morphology and photosynthetic activity were allelic to the causal gene for narrow leaf mutant in rice, suggesting the availability of narrow leaf genes for breeding high-yield varieties [19–23].

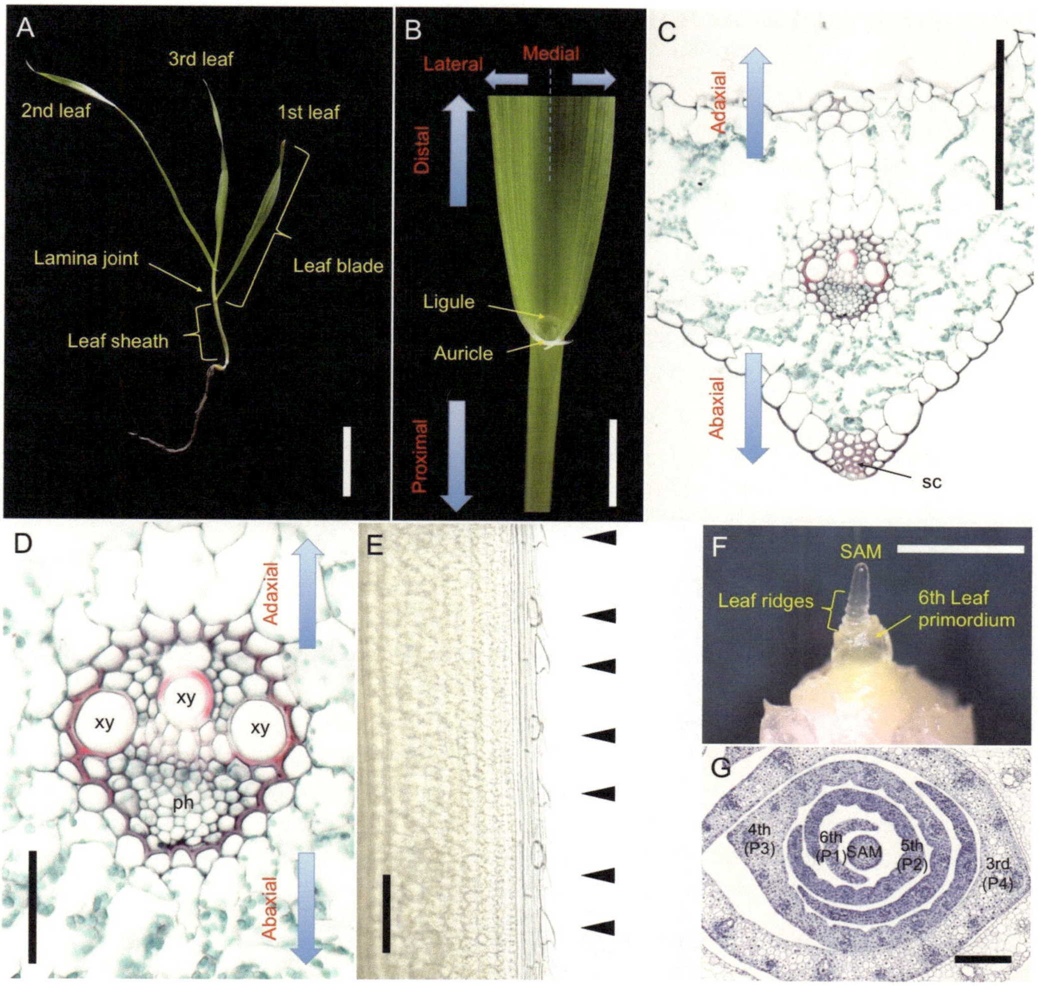

Figure 1. The shoot structure of normal barley (KN29). **(A)** A barley seedling at the second leaf stage. The leaf stage is defined by the number of fully expanded leaves. The first to third leaves are labeled. The leaf blade, leaf sheath, and lamina joint in the first leaf are indicated. **(B)** Close-up of the lamina joint in the second leaf. The ligule and auricle are pointed by arrows. **(C)** A cross section of the medial region in the second leaf blade. The section is double-stained in safranin and fast green. The lignified tissue is stained in red by safranin. "sc" indicates sclerenchymatous cells. **(D)** Close-up of the central vascular bundle in **(C)**. "xy" and "ph" indicate xylems and phloems, respectively. **(E)** The epidermal cells of the leaf margin in the second leaf blade. Arrow heads indicate the sawtooth hairs in the leaf margin. **(F)** A shoot apex of barley seedling at the second leaf stage. Matured leaves and leaf primordia are removed. The sixth leaf is initiating from the basal part of shoot apical meristem (SAM). Barley is unique in that leaf ridge formation precedes leaf primordium development. **(G)** A cross section of the shoot apex in barley seedling at the second leaf stage. The leaf positions (third to sixth) and the leaf primordial stages (P1–P4) are shown in the figure. Bars = 5 cm **(A)**, 5 mm **(B)**, 200 μm **(C, G)**, 50 μm **(D)**, 100 μm **(E)**, 1 mm **(F)**.

Here, the molecular mechanisms underlying the leaf development in medial-lateral direction are discussed as we introduce the three major groups of narrow leaf mutants in grass family: (1) auxin-related mutants, (2) cellulose synthase-like D (CSLD)-related mutants, and (3) polarity-related mutants.

2. Auxin-related narrow leaf mutants

Auxin is a fundamental plant hormone and regulates a variety of plant growth and development. All parts of the young plant such as cotyledons, expanding leaves, and root tissues can potentially produce auxin although the youngest leaves exhibit the highest biosynthetic capacity [24–26]. Auxin is unique in its polar transportation (polar auxin transport (PAT)), as we mentioned above, mediated by influx carriers and efflux carriers [7]. The direction of auxin flow is the consequence of asymmetric localization of these carriers at plasma membrane [27, 28]. The resulting auxin localization within organs plays pivotal roles in phyllotactic patterning [29, 30], organogenesis [9, 31, 32], embryogenesis [33, 34], tropic response [35], and apical dominance [36]. At the cellular level, auxin regulates cell division, cell elongation, and cell differentiation [7, 37].

The predominant form of auxin is indole-3-acetic acid (IAA). Genetic and biochemical analyses indicated that tryptophan (Trp) is the main precursor of IAA in plants, and four biosynthetic pathways for IAA from Trp have been assumed [38–40]. Among IAA biosynthetic enzymes revealed so far, the most important biosynthetic enzymes are the tryptophan aminotransferase of arabidopsis (TAA) family of aminotransferases and the YUCCA (YUC) family of flavin-containing monooxygenases [41, 42]. TAA1 catalyzes the conversion of Trp to indole-3-pyruvic acid (IPA) in the initial step of the IPA pathway, and YUC catalyzes the conversion of IPA to IAA, downstream of TAA, in *arabidopsis* [40, 42–45]. The inactivation of a single *TAA* or *YUC* gene showed no obvious defects, indicating overlapping functions among *TAA* or *YUC* family members. On the other hand, the simultaneous inactivation of *TAA1* and its close homologs, *TAA-related1* (*TAR1*) or *TAR2* (**Figure 2A**), or inactivation of two or more *YUC* genes resulted in multiple growth defects together with a severe reduction in IAA level [43, 46]. Therefore, the IPA pathway, catalyzed by TAA and YUC, is considered to be the major auxin biosynthetic pathway in *Arabidopsis* [40].

The importance of the IPA pathway in IAA biosynthesis is also demonstrated in grass family. In maize, loss-of-function of *vanishing tassel2* (*VT2*) and *sparse inflorescence1* (*SPI1*), co-ortholog of *TAA1* and *YUC* in maize, respectively (**Figure 2**), caused severe barren inflorescences and semidwarf vegetative phenotypes with fewer leaves together with the reduction in IAA content [47, 48]. Similar reduction in IAA levels was shown in the loss-of-function of *fish bone* (*FIB*) and *narrow leaf7* (*NAL7*), co-ortholog of *TAA1* and *YUC* in rice, respectively (**Figure 2**) [49, 50]. Thus, the IPA pathway seems to be the major IAA biosynthetic pathway in plants.

The reduction in IAA levels gives rise to pleiotropic organ malformation together with severe narrow leaf phenotype in rice. *Tryptophan deficient dwarf1* (*TDD1*) encodes a protein homologous to the anthranilate synthase β-subunit, which catalyzes the initial step of the Trp biosynthesis pathway [51]. *TDD1*mutant is embryonic lethal because of a failure to develop most organs during embryogenesis. Regenerated *TDD1* plants exhibit pleiotropic malformations including dwarfing, narrow leaves, short roots, and abnormal flowers, together with a reduction in Trp and IAA content. Trp feeding and moderate expression of *OsYUC1* rescued the mutant phenotypes, indicating that abnormal phenotypes of *TDD1* were caused mainly by Trp and IAA deficiency [51]. The loss-of-function of *constitutively wilted 1* (*COW1*), which encodes

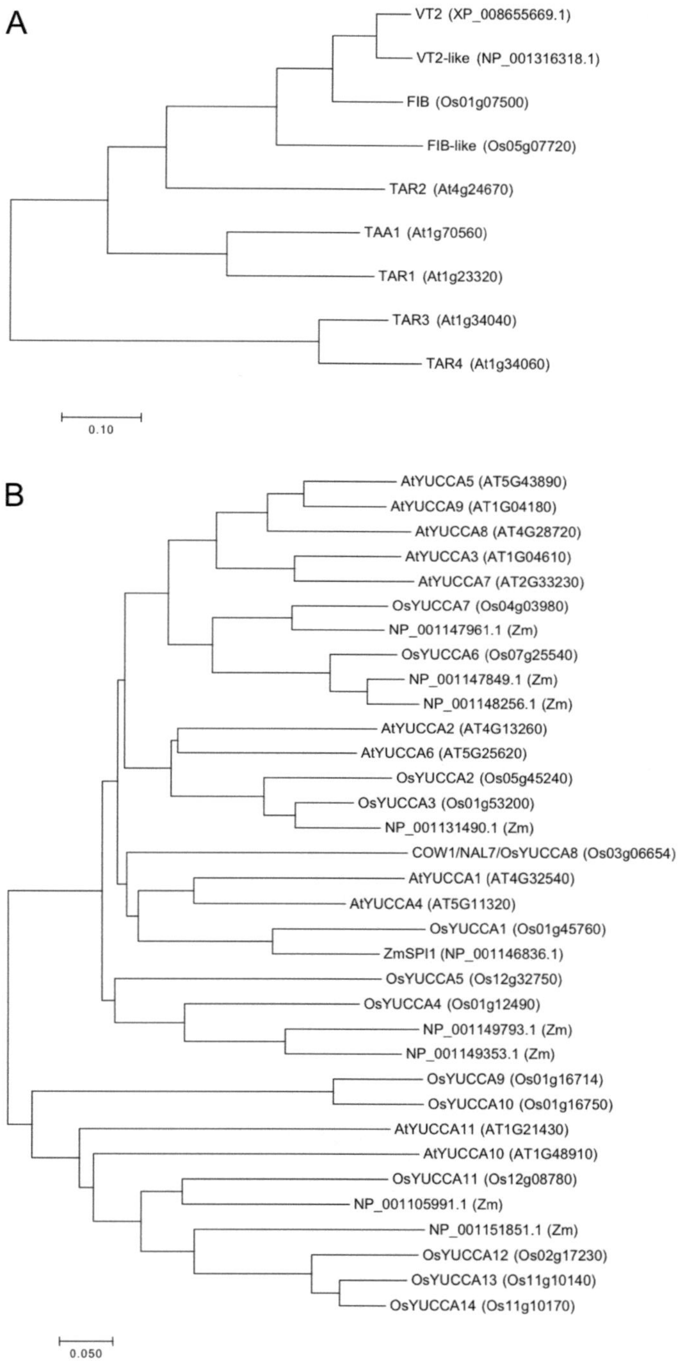

Figure 2. Phylogenetic tree of proteins involved in the IPA pathway. **(A)** TAA-related proteins in rice (FIB and FIB-like), maize (VT2 and VT2-like), and *Arabidopsis thaliana* (TAA1 and TAR1-4). **(B)** YUCCA-related proteins in rice (Os), maize (Zm), and *Arabidopsis thaliana* (At).

OsYUC8 (**Figure 2B**), was isolated from *TOS17* and T-DNA insertional rice mutants [52]. *COW1*mutants exhibited narrow leaves and a rolled leaf phenotype, which is likely attributable to insufficient water supply due to the small root-to-shoot ratio. Fujino et al. [49] identified another allele of *COW1*, *narrow leaf 7* (*NAL7*). The *NAL7* mutant shows a similar but milder phenotype compared with *COW1*, and the IAA content in *NAL7* was reduced compared to the wild type. In addition, overexpression of *NAL7* cDNA gave rise to overgrowth and abnormal morphology of the root, which was likely attributable to the overproduction of auxin. These results suggested that *NAL7/OsYUC8* is also involved in auxin biosynthesis. The importance of *TAA* gene in IAA biosynthesis in rice was demonstrated by *fish bone* (*FIB*) mutant [50]. *FIB* exhibited pleiotropic abnormal phenotypes including dwarfing, narrow and adaxially rolled leaves with large lamina joint angles, abnormal vascular development, and lack of crown and lateral roots. In addition, *FIB* also showed lack of gravitropism and aberrant phyllotaxy deviated from the normal distichous one. Map-based cloning revealed that *FIB* encodes co-ortholog of *TAA1* in rice (**Figure 2**A). Interestingly, loss-of-function of *FIB* resulted in not only the reduction in IAA level but also higher sensitivity to IAA and lower PAT activity. These results suggest that auxin biosynthesis, transport, and sensitivity are interrelated, which might be attributable to the pleiotropic abnormal phenotypes of *FIB* [50]. Rice genome includes 2 and 14 genes belong to the *TAA* and *YUC* families, respectively (**Figure 2**) [53]. While the inactivation of a single *TAA* or *YUC* gene showed no obvious defects in *arabidopsis*, distinct abnormal phenotypes were appeared in *FIB* or *COW1/NAL7* mutants in rice, suggesting that functional redundancy among *TAA* or *YUC* genes is less prevalent in rice than in *Arabidopsis*.

In contrast, rice *narrow leaf1* (*NAL1*) encodes a trypsin-like serine and cysteine protease, whose relationship between auxin remains unknown, but *NAL1* mutant showed narrow leaves, dwarfing, and defective vascular patterns together with reduced PAT activity [54]. Surprisingly, several agronomic QTLs involved in flag leaf width (*qFLW4*; [19], *WFL*; [23]), photosynthesis rate (*GPS*; [21]), flag leaf shape (*qLSCHL4*; [22]), and spikelet number (*SPIKE*; [20]) were allelic to *NAL1*. The increased yield in *indica* rice varieties, which introduced these QTLs, suggests that *NAL1* is available in plant breeding. The latest study uncovered that *NAL1* functions in the regulation of cell division during leaf primordia initiation [55]. In *NAL1* mutant, expression of several G1- and S-phase specific genes were reduced, suggesting that *NAL1* affects cell-cycle regulation. In addition, the reduced expressions were also shown in *PIN1*, three *auxin response factor ARF* genes, and three *YAB* genes, but the expression of *YUC* genes were comparable to those of wild type. These results indicated that the inactivation of *NAL1* affects auxin transport and auxin response but not auxin biosynthesis [55].

Overall, auxin-related narrow leaf mutants exhibit pleiotropic abnormal phenotypes other than the reduction in leaf width. The representative phenotypes seem to be appeared in vascular patterning and root growth since auxin plays critical role in the development of these organs.

3. CSLD-related narrow leaf mutants

Cell walls are essential structures surrounding plant cells. While cells are expanding, primary cell walls fulfill the support and barrier functions. After cell expansions are completed, secondary cell walls are formed between primary walls and plasma membranes, giving additional strength to cells. Cell wall is composed of polysaccharides, proteins, and phenolic compounds. Classically, polysaccharides are classified into cellulose, hemicelluloses, and pectins [56]. Cellulose synthase (CesA) protein contains a zinc finger domain at the *N*-terminus, eight transmembrane domains, and a central catalytic domain known as "D_D_D_QxxRW" motif. Although the mechanism by which CesA creates β-1,4-glucan chain is not fully revealed, it is plausible that glucan chain synthesized by the catalytic domain in the cytoplasm goes out of plasma membrane through the pores formed by the transmembrane domain [57]. It is likely that the zinc finger domain at the *N*-terminus is involved in CesA protein dimerization, leading to the higher-order structures [58, 59].

Based on the sequence similarity to *CesA* genes, a large superfamily of at least 41 *cellulose synthase-like* (*CSL*) genes were found in the *Arabidopsis thaliana* genome [60]. They were classified into six subfamilies (*CSLA, B, C, D, E,* and *G*), and subsequent studies identified three additional *CSL* subfamilies (*CSLF, H,* and *J*) [61, 62]. CSL proteins contain sequence motifs that are characteristics of β-glycosyltransferases. The only difference of CSLs from CesAs is the lack of the zinc finger domains at the *N*-terminus, which seems to be particularly important to form higher-order structures. In addition, most CSL proteins appear to be localized not in the plasma membrane but in the Golgi, where hemicellulose synthesis takes place. From these characteristics, *CSL* genes are predicted to catalyze the biosynthesis of noncellulosic polysaccharides [60]. As far as we know, the first biochemical evidence was provided by the soybean somatic embryos, in which expression of guar *CSLA* candidate cDNA gave rise to the enhanced mannan synthase activity [63]. Subsequent studies demonstrated that the *CSLA* genes encode (gluco)mannan synthases [64, 65], and that the *CSLF* and *CSLH* genes encode mixed linkage glucan synthases [66, 67]. *CSLC* genes were predicted to be involved in the xyloglucan synthesis [68], but recent study reported that some *CSLC* genes of barley are targeted to the plasma membrane, suggesting that the *CSLC* subfamily contains more than one type of polysaccharide synthase [69].

The uneven distribution of *CSL* genes implies how *CSL* subfamilies have been evolved in parallel with the diversification of plant species. While *CSLB* and *CSLG* subfamilies are found only in eudicots, *CSLF, CSLH,* and *CSLJ* subfamilies are specific to Poaceae. Particularly, *CSLJ* subfamily is unique in that it is only found in certain grasses, such as barley, wheat, sorghum, and maize, but not in rice or *Brachypodium* [62]. In contrast, *CSLD* subfamily is commonly found in all land plants, and show the highest similarity to *CesA* family among *CSL* subfamilies at sequence levels. The small number of introns and the gene structure diversity within the subfamily imply the possibility that *CSLD* is the oldest gene family in the cellulose synthase superfamily [60, 70]. Genome database survey revealed that *CSLD* subfamily contains six *Arabidopsis* genes, five maize genes, five rice genes, three barley genes, five sorghum genes, and six *Brachypodium* genes, and subsequent phylogenetic analysis showed that they are further classified into three clades (**Figure 3**) [71, 72]. The first clade including *AtCSLD1* and *AtCSLD4* is specifically expressed in pollens and involved in pollen tube elongation [73], and the second clade including *AtCSLD2, AtCSLD3, OsCSLD1,* and *ZmCSLD5* is highly expressed in root tissues and involved in root hair development [73–77]. While these two clades are commonly involved in "tip-growing" development, the loss-of-function of the third clade including *AtCSLD5, OsCSLD4,* and *ZmCSLD1* exhibited different phenotypes.

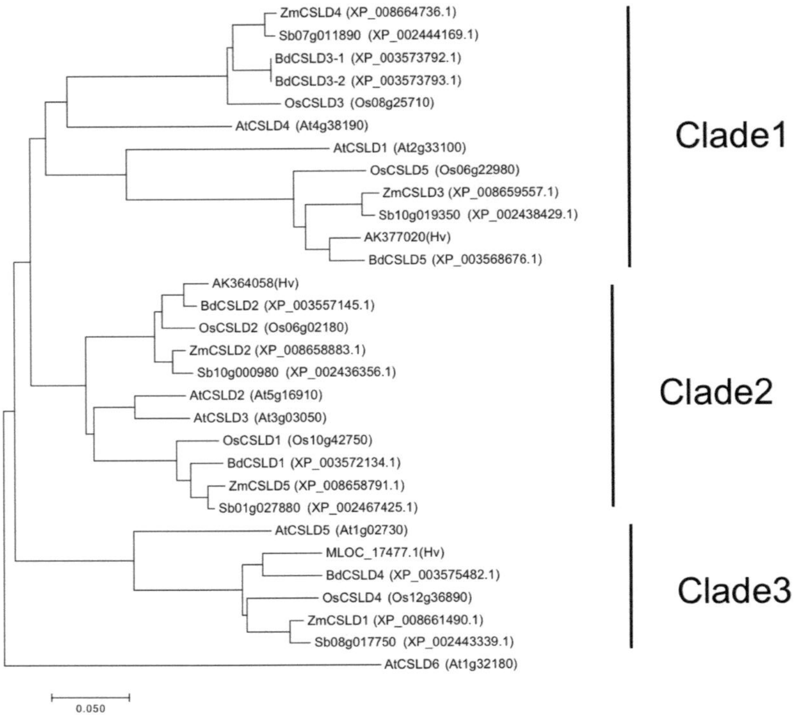

Figure 3. Phylogenetic tree of CSLD-related proteins in rice (Os), maize (Zm), barley (Hv), sorghum (Sb), *Brachypodium distachyon* (Bd), and *Arabidopsis thaliana* (At).

In rice, inactivation of *OsCSLD4* resulted in distinct narrow leaf phenotype. So far, several narrow leaf genes such as *narrow leaf and dwarf 1* [78], *narrow and rolled leaf 1* [79], *Oscd1* [80], *slender leaf 1* [72], *dwarf and narrowed leaf 1* [81], and *dwarf and narrow leaf 3* [82] were allelic to *OsCSLD4*. The mutants commonly exhibited narrow and rolled leaves and dwarfing phenotypes. The reduction in leaf-blade width and plant height was clearly attributable to the decrease of cell number, suggesting that *OsCSLD4* promotes cell proliferation activity. But if so, why is leaf-blade length less affected by the mutation than that of leaf-blade width? This question was solved by the increase of cell length in *OsCSLD4* mutant. Plants are able to compensate for a reduction in cell number with an increase in cell size [83], and the degree of compensation may differ depending on the direction. In fact, the number of cells was equally reduced in both length and width direction in *OsCSLD4* [72]. The expression analysis revealed that *OsCSLD4* is specifically expressed in M-phase cells in all developing organs, and loss-of-function of *OsCSLD4* resulted in the alteration of cell-cycle regulation. Interestingly, *OsCSLD4* included cells with 4C nucleus while such cells were not detected in normal rice. These results suggested that *OsCSLD4* plays a pivotal role in *M*-phase to progress cell proliferation [72].

The inactivation of *ZmCSLD1* also results in the narrow leaf and fine stem phenotype mainly due to the decrease of cell number [71]. In addition, wart-like cell clusters were formed on the leaf surface. The warts were attributable to the defects of cell division in leaf development, and disrupted cross-wall formations were frequently observed in epidermal cells. Such defective developments of cell wall often appeared in cytokinetic mutants of *Arabidopsis*, such as *knolle* [84], *korrigan* [85], and *hinkel* [86], in which impairment of cytokinesis was caused by a failure of cell-plate formation. Considering the nature of *CSLD* as a wall-synthesizing enzyme, the *M*-phase specific expression, and the defective cell wall development in the mutant, it is speculated that *CSLD* may be involved in cell-plate formation. The existence of *CSLD* genes in all land plants also suggests the fundamental function of this subfamily. Recently, it was revealed that transiently expressed *AtCSLD5* is involved in mannan synthesis in tobacco leaves [87]. Distantly related *CSLA* subfamily also exhibits mannan synthase activity, but CSLA proteins readily use Guanosine diphosphate (GDP)-glucose as well as GDP-mannose and hence efficiently synthesize glucomannans [64, 88]. Since the mannosyltransferase activity of *AtCSLD5* was reduced by adding GDP-glucose together with GDP-mannose, *CSLD* subfamily is involved in a different kind of mannan synthesis from that catalyzed by *CSLA* subfamily [87]. Although mannans have been well studied as storage component, little information has been accumulated in relation to cytokinesis. Further analysis will reveal the detailed mechanism of plant cytokinesis and novel functions of hemicelluloses.

Overall, *CSLD*-related narrow leaf mutants exhibit a decrease in the whole plant size other than the reduction in leaf width. These phenotypes are directly attributable to the reduced cell proliferation activity, for CSLDs of the third clade are predicted to fulfill a function closely related to cytokinesis.

4. Polarity-related narrow leaf mutants

Most plant leaves are asymmetrical in all directions. Grass family leaves include leaf-blade in the distal side, leaf-sheath in the proximal side, and lamina-joint between the leaf-blade and leaf-sheath. The bulliform cells, which curl leaf-blades to prevent over transpiration, and xylems are formed only on the adaxial side, and the phloems on the abaxial side. The midrib, which functions as a physical support for the leaves, and ligule are formed in the medial side, and the sawtooth hairs and auricle in the lateral side (**Figure 1A–E**) [89]. For the construction of such a sophisticated organ, the proximal-distal, adaxial-abaxial, and medial-lateral polarities must be constructed as soon as cells acquire leaf fate in SAM (**Figures 1F** and **G**).

Among the three polarities, the molecular mechanism of adaxial-abaxial polarity is well studied using *Arabidopsis*. Through the loss-of-function and/or gain-of-function analyses, it has been revealed that adaxial identity is regulated by class III homeodomain-leucine zipper (HD-Zip) family genes and *asymmetric leaves2* (*AS2*), and that abaxial identity by *yabby* (*YAB*) family genes, *kanadi* (*KAN*) family genes, and *auxin response factor* (*ARF*) family genes [90]. The adaxial or abaxial specific expression of these genes is crucial for the establishment of the organ polarity, and these regulators are interacting antagonistically [17, 90]. In addition, small RNAs are also involved in the negative regulation of these regulators to maintain the expression regions [90–92]. In rice, the loss-of-function of *shallot-like 1* (*SLL1*)/*rolled leaf 9* (*RL9*), which encodes SHAQKYF class MYB transcription factor belonging to the *KAN* family, resulted in the suppression of abaxial development while enhanced expression led to the abaxialized leaf phenotypes [93, 94]. Moreover, in maize, the accumulation of miR166, which is involved in the cleavage of class III *HD-Zip* transcripts, defined the expression region of the *rolled leaf 1* (*RLD1*) belonging to the class III *HD-Zip* family, promoting the establishment of adaxial-abaxial polarity [95]. Despite the morphological differences from dicots, these genes homologous to *KAN* or class III *HD-Zip* seem to fulfill similar regulation in grass family.

While detail genetic regulators of proximal-distal polarity remain unclear in *Arabidopsis*, morphological, and molecular analyses are proceeding in grass families for the convenience of distinct organ development along the proximal-distal axis. A number of dominant mutations which specifically affect proximal-distal patterning have been characterized in maize [96]. The dominant mutant *Knotted1* (*Kn1*) was characterized by sheath-like cells in the leaf-blade [97]. *KN1* encodes a homeodomain protein, and *KN1* transcripts were localized in the meristem but excluded from the leaf initial cells [98, 99]. However, KN1 proteins were detected outside of the *KN1*-transcript localized area, suggesting the noncell-autonomous nature of *KN1* gene [100]. In leaf primordia, KN1 proteins were accumulated in the most proximal part, and ectopic expression of *KN1* in the distal leaf-blade gave rise to alteration into sheath cell identity [101, 102]. These results suggested that *KN1* is involved in the establishment of proximal identity in leaf development. Ectopic expressions of *KN1-like homeobox* (*KNOX*) genes also resulted in cell fate alterations in maize, barley, and *Arabidopsis*, suggesting the highly conserved function of *KNOX* genes [103–105]. On the other hand, PIN1 proteins which mediate polar auxin transport (PAT) are highly expressed in the distal ends of developing leaf primordia. Auxin plays pivotal roles in leaf development as we mentioned above, and PIN1 creates an auxin maximum in the distal end of leaf primordium [31, 106]. The subsequent canalization through the interior of leaf primordia leads to the development of primary vascular strand. Thus, the auxin gradient along the proximal-distal axis is likely to play pivotal role in leaf development. Maize *liguleless1* (*LG1*) and *liguleless2* (*LG2*) mutants lack both ligule and auricle between leaf-blade and leaf-sheath [107–109]. It was revealed that *LG1* encodes a squamosa-promoter binding protein, and that *LG2* encodes a basic leucine zipper protein [110, 111]. While *LG1* is specifically expressed in ligule initiating area, *LG2* shows earlier and broader expression pattern than that of *LG1* [112, 113]. The phenotype of *lg1 lg2* double mutant suggested that they act in the same pathway, implying the possibility of interaction between *LG1* and *LG2* [109]. In addition, other liguleless mutants have been identified such as *LG3* and *LG4*, which encode class I *KNOX* genes [114, 115]. These findings promote the construction of a hypothetical model of leaf-blade-sheath boundary formation [113].

Compared with other polarities, the molecular mechanism of the medial-lateral polarity is less understood. So far, it was revealed that *drooping leaf* (*DL*) plays pivotal role in the development of medial organs in rice. *DL* encodes a putative transcription factor belonging to the *YAB* family, and *DL* mutants showed defective development of a midrib in the leaf, leading to the drooped leaf phenotype [116]. The *DL* transcripts were localized in the central region of leaf primordia, and over-expression of *DL* resulted in the ectopic formation of midrib-like structures in the lateral regions as well as in the central region of the leaf. In contrast, the development of leaf lateral domains is highly regulated by *wuschel-related homeobox* (*WOX*) genes. In maize, the loss-of-function mutations in both *narrow sheath1* (*NS1*) and *NS2* resulted in the significant reduction in leaf width due to the lack of marginal regions in leaves [117–119]. *NS1 and NS2* double mutants fail to downregulate KNOX proteins in the premarginal regions of leaf primordia, leading to the deletion of marginal region from the primordial stages [117, 119]. *NS1* and *NS2* encode the duplicated *WOX3* genes, and *NS* transcripts are accumulated in the marginal edges of initiating leaf primordia. From these results, it was suggested that *NS* genes play pivotal roles in the recruitment of leaf founder-cells by downregulating KNOX accumulation [120–122]. Genes belonging to *WOX3* family are largely classified into two clades (**Figure 4**), and the *NS*-related clade includes *narrow leaf2* (*NAL2*) and *NAL3* of rice, *narrow leafed dwarf 1* (*NLD1*) of barley, and *pressed flower1* (*PRS1*) of *Arabidopsis* [123].

The nucleotide sequences of *NAL2* and *NAL3* are identical, corresponding with the recent duplication of a large chromosomal segment in chromosomes 11 and 12 [124]. *NAL2 /3* and *NLD1* mutants show the similar abnormal phenotypes to *NS1* and *NS2* such as distinct narrow leaf phenotype and defective marginal development, which are attributable to the lack of marginal regions (**Figure 5**) [123, 125, 126]. The expression patterns of *NAL2/3* and *NLD1* are also similar to that of *NS1 NS2*, suggesting the conserved function of *NS*-related genes in the development of lateral organs. Interestingly, no distinct abnormal phenotypes were observed in the leaf of *PRS1* mutant except for the deletion of the proximal lateral stipules [117]. This result supported the leaf-zonation model that the lower leaf zone of bifacial monocot leaves corresponds with the basal part of bifacial eudicot leaves including stipules [127]. It is, therefore, considered that *NS*-related *WOX3* genes are involved in the development of the lateral domain in the lower leaf zone.

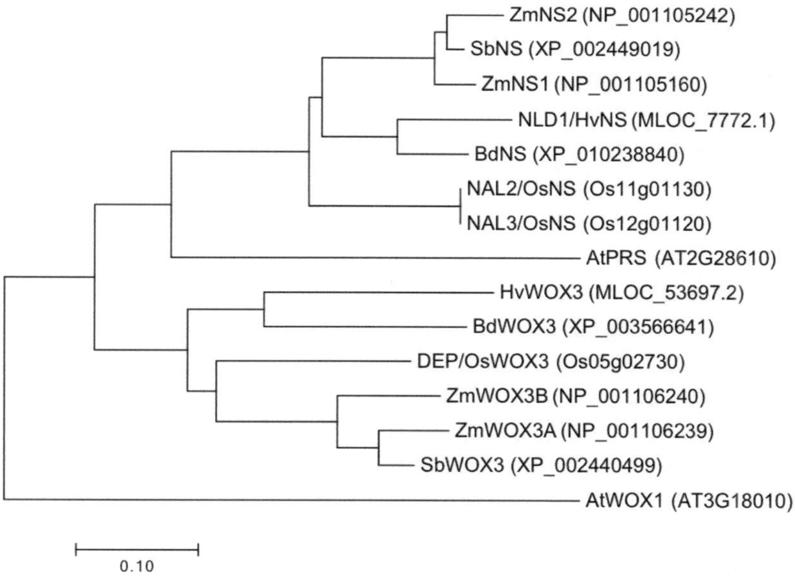

Figure 4. Phylogenetic tree of WOX3-related proteins in rice (Os), maize (Zm), barley (Hv), sorghum (Sb), *Brachypodium distachyon* (Bd), and *Arabidopsis thaliana* (At).

Figure 5. The narrow leaf phenotype of barley *narrow leafed dwarf1* (*NLD1*) mutant. **(A–C)** The whole shoots **(A)**, the leaf-blades **(B)**, and the lamina joints **(C)** in matured leaves of wild-type (KN29) and *NLD1.b* mutant. The auricles are pointed by arrows in **(C)**. Auricles are significantly diminished in *NLD1.b* due to the defective development of the lateral domain. **(D)** The epidermal cells of the leaf margin in *NLD1.b* leaf-blade. Sawtooth hairs are rarely formed in the mutant unlike wild-type (**Figure 1E**). Bars = 1 cm **(B, C)**, 200 μm **(D)**.

The width of *PRS* leaves was significantly reduced by the additional mutation of *WOX1*. *WOX1* is unique in that it belongs to the same clade of the *WOX3/PRS* family but seems to be absent in grasses (**Figure 4**) [128, 129]. *WOX1* and *PRS* double mutants exhibit not only the loss of leaf marginal tissues but also the confused adaxial-abaxial identity at leaf marginal regions [129, 130]. These results suggested that leaf margin functions as an adaxial-abaxial boundary, where adaxial and abaxial regulators are downregulated by *WOX* genes [90].

While the medial-lateral polarity is directly related to leaf width, mutation or over-expression of the genes regulating the proximal-distal or adaxial-abaxial polarity can also result in the reduction in leaf width together with the alteration of organ polarities. Recessive mutant *rough sheath 2* (*RS2*) of maize exhibits narrow/bladeless leaves with a disruption of the blade-sheath boundary [131]. In *RS2* mutant, class I KNOX proteins are ectopically accumulated, and it was revealed that *RS2* encodes an MYB-domain protein, an ortholog of *AS1* in *Arabidopsis*. Thus, it is likely that *RS2* is involved in the proximal-distal patterning by downregulating *KNOX* expression. *Liguleless* (*LG*) genes play pivotal role in the establishment of the boundary between leaf-blade and leaf-sheath in the proximal-distal axis. Recently, another *LG* gene *liguleless narrow* (*LGN*) was identified, and its semidominant mutant (*LGN-R*) showed narrow leaves with greatly reduced auricle and ligule and indefinite blade-sheath boundary [132]. *LGN* encodes a grass-specific kinase, which is broadly expressed in maize organ but affects *LG1* and *LG2* expression. The dominant mutant *Wavy auricle in blade 1* (*WAB1*) shows narrow leaves with ectopic auricle and extended sheath in leaf-blade [112, 133]. In contrast to *LGN-R*, *LG1* was misexpressed in *WAB1*, and recently it was revealed that *WAB1* encodes a teosinte-branched1/cycloidea/PCF (TCP) transcription factor, which is necessary for *LG1* expression [134]. These genes play a pivotal role in the establishment of proximal-distal polarity, but affect leaf width indirectly. Thus, it was considered that proximal-distal patterning may link to medial-lateral growth.

On the other hand, Rice *SLL1/RL9* encodes KAN transcription factor as we mentioned above, and *sll1/rl9* mutants show rolled leaf phenotypes due to the defective development of the sclerenchymatous cells on the abaxial side together with the reduction in leaf width [93, 94]. Similar defective development was observed in *semi-rolled leaf 2* (*SRL2*), which exhibits narrow incurved leaves due to the defective development of sclerenchymatous cells on the abaxial side [135]. *SRL2* encodes a novel plant-specific protein of unknown biochemical function, and highly expressed in the abaxial cell layer in the leaf sheath. However, *SLL1/RL9* expression was unaffected in *SRL2*, and *SRL2 SLL1* double mutants showed more severe defective development of sclerenchymatous cells on the abaxial side together with the much narrower leaf phenotype than single mutants [135]. These results suggest that *SLL1/RL9* and *SRL2* function in distinct pathways to regulate the abaxial development. Overexpression of *OsHOX32*, a member of class III *HD-Zip* family, resulted in narrow and adaxially rolled leaves due to the reduction in bulliform cell number [136]. Among the six *OsYAB* genes, *OsYAB1*, *OsYAB2*, and *OsYAB6* were upregulated while *OsYAB3*, *OsYAB4*, and *OsYAB5* were downregulated in the overexpression plants, suggesting the direct or indirect interaction between *OsHOX32* and *OsYAB* genes. Similar defective development was observed by the overexpression of *OsLBD3-7*, which shows high similarity to *AS2* of *Arabidopsis*. *OsLBD3-7* overexpression plants exhibit narrow and adaxially rolled leaves due to the reduction in bulliform cell size and number [137]. Since the negative regulators of bulliform cell development were upregulated in overexpression plants, it was suggested that *OsLBD3-7* positively regulate these negative regulators in leaf development. The marginal expressions of *NS* genes are disappeared in maize *ragged seedling 2* (*RGD2*) mutant, which exhibits thread-like narrow leaves [138]. *RGD2* encodes argonaute7 (AGO7)-like protein, which is involved in the synthesis of trans-acting short-interfering RNA (ta-siRNA) derived from *TAS3* in *Arabidopsis*. So far, several mutants for *TAS3* ta-siRNA pathway have been identified including *AGO7*-related genes (*RGD2* in maize; [138], *shootless4* (*SHL4*)/*shoot organization2* (*SHO2*) in rice; [139]), *SGS3*-related gene (*leafbladeless1* [*LBL1*] in maize; [140, 141]), *RDR6*-related gene (*SHL2* in rice; [142, 143]), and *DCL4*-related gene (*SHO1* in rice; [143, 144]). Although maize and rice leaves are different morphologically, the loss-of-function of these genes commonly gave rise to thread-like narrow leaves which showed defective adaxial-abaxial and medial-lateral polarities. *TAS3* ta-siRNA is expressed on the adaxial side of developing leaf primordia and restricts the expression region of abaxial factor *ARF3a* and miR166 [141]. Since miR166 restricts the expression region of class III *HD-Zip* genes, inactivation of *TAS3* ta-siRNA pathway results in the upregulation of *ARF3a* and miR166, and downregulation of class III *HD-Zip* genes, leading to the abaxialization of leaf. Such a severe abaxialization might disturb the establishment of medial-lateral polarity. In *Arabidopsis*, triple mutation of *YAB* genes (*FIL YAB3 YAB5*) has resulted in the thread-like narrow leaves which showed defective adaxial-abaxial and medial-lateral polarities [145, 146]. These results suggest that the establishment and/or development of the medial-lateral polarity is regulated downstream of the adaxial-abaxial polarity.

Overall, polarity-related narrow leaf mutants exhibit distinct reduction in leaf-blade width together with the disruption of organ polarity. The loss-of-function of lateral identity is directly reflected in the reduction of leaf width, but the disruption of the proximal-distal or adaxial-abaxial polarities also affect the establishment or development along medial-lateral axis, suggesting the interactive development between the three polarities.

5. Conclusion

The reduction in leaf width is a subtle morphological alteration, but the analyses of narrow leaf mutants have uncovered molecular functional diversity of the causal genes. Through a variety of genetic approaches, it has been demonstrated that *NS*-related *WOX3* genes are critical for the development of leaf lateral domains. Although *NS*-related *WOX3* transcripts are strictly limited within the marginal edges, the phenotypic alteration of loss-of-function mutants occurs in more broad area, suggesting the noncell-autonomous nature of *NS*-related *WOX3* genes. This could be explained by the migration of either WOX3 protein itself or the secondary signals derived from the marginal cells. Recently, it was reported that barley *NLD1* mutant exhibited malformation of commissural veins in the leaf lateral domain [123]. Since polar auxin transport plays an important role in determining vascular pattern in leaves, *nld1* may include some abnormalities in auxin transport. Therefore, it is quite interesting whether auxin functions as the secondary signal of *NS*-related *WOX3* genes. Auxin plays pleiotropic role in plant development, and at the cellular level, auxin regulates cell division, cell elongation, and cell differentiation. In addition, it is suggested that auxin biosynthetic *YUC* genes are expressed in response to the juxtaposition of adaxial and abaxial domains [147]. Thus, auxin biosynthesis at the adaxial-abaxial boundary partly contributes to leaf margin expansion, and this might explain the reduction in leaf width attributable to the disruption of the adaxial-abaxial polarity. At the downstream of these mechanisms, cell proliferation activity is maintained by *CSLD* genes of the third clade. The details of plant cytokinesis are not fully understood, particularly as to the components of cell plate. All we covered here is just a part of well-studied mutants, and there should be many hither-to unidentified narrow leaf mutants. Further study will give us a novel and detailed mechanism of leaf development in the grass family.

6. Materials and methods

6.1. Plant materials

For morphological observation of barley shoot, a wild type line Kanto Nijo 29 (KN29), which has two-rowed spike and covered caryopsis, and its gamma-ray induced *narrow leafed dwarf1* (*NLD1*) mutant, *NLD1.b*, were used. To promote germination, seeds were kept at 15°C on wet paper for 3 days. Then, imbibed seeds were sown on soil and grown under natural conditions.

6.2. Paraffin sectioning and histological analysis

Plant samples were fixed with FAA (formaldehyde:glacial acetic acid:50% ethanol [2:1:17]) for 24 h at 4°C for histological analysis. They were then dehydrated in a graded ethanol series, substituted with 1-butanol, and embedded in Paraplast® Plus (McCormick Scientific). The samples were sectioned at 8 μm thickness using a rotary microtome. For the histological analysis, sections were stained in hematoxylin or double-stained in safranin and fast green. After staining, sections were mounted with Poly-Mount® (Polysciences, Inc.) and observed with a light microscope.

6.3. Epidermal cell observation

The leaf-blades were fixed with FAA (formaldehyde:glacial acetic acid:50% ethanol [2:1:17]) for 24h at 4°C. They were then dehydrated in a graded ethanol series. Dehydrated samples were incubated at 96°C in chloralhydrate dissolved in 100% ethanol until they were cleared, and observed with a light microscope.

6.4. Phylogenetic analysis

For the phylogenetic analysis of *TAA-*, *YUCCA-*, *CSLD-*, and *WOX3*-related genes, amino acid sequences were obtained from TIGR (http://rice.plantbiology.msu.edu) for rice, IPK Barley BLAST Server (http://webblast.ipk-gatersleben.de/barley/) for barley, NCBI (https://www.ncbi.nlm.nih.gov) for maize, sorghum, and *Brachypodium distachyon*, and TAIR (https://www.arabidopsis.org) for *Arabidopsis thaliana*. As for YUCCA-related maize proteins, amino acid sequences showing the highest similarity to YUCCA protein were searched using the protein blast in NCBI. The obtained sequences were analyzed with MEGA version 7 (available at http://www.megasoftware.net, [148]) to create the phylogenetic trees.

Author details

Takanori Yoshikawa[1*] and Shin Taketa[2]

*Address all correspondence to: t-yoshi@kiui.ac.jp

1 School of Agricultural Regional Vitalization, Kibi International University, Minamiawaji, Japan

2 Group of Genetic Resources and Functions, Institute of Plant Science and Resources, Okayama University, Kurashiki, Japan

References

[1] Falster DS, Westoby M. Leaf size and angle vary widely across species: What consequences for light interception?. New Phytologist. 2003;**158**:509-525

[2] Hay A, Craft J, Tsiantis M. Plant hormones and homeoboxes: Bringing the gap?. BioEssays. 2004;**26**:395-404

[3] Fleming AJ. Formation of primordia and phyllotaxy. Current opinion in Plant Biology. 2005;**8**:53-58

[4] Hake S, Smith HMS, Holtan H, Magnani E, Mele G, Ramirez J. The role of KNOX genes in plant development. Annual Review of Cell and Developmental Biology. 2004;**20**:125-151

[5] Hay A, Tsiantis M. KNOX genes: Versatile regulators of plant development and diversity. Development. 2010;**137**:3153-3165

[6] Hepworth SR, Pautot VA. Beyond the divide: Boundaries for patterning and stem cell regulation in plants. Frontiers in Plant Science. 2015;**6**:1052

[7] Petrášek J, Friml J. Auxin transport routes in plant development. Development. 2009;**136**: 2675-2688

[8] Moon J, Hake S. How a leaf gets its shape. Current Opinion in Plant Biology. 2011;**14**:24-30

[9] Křeček P, Skůpa P, Libus J, Naramoto S, Tejos R, Friml J, Zažímalová E. The PIN-FORMED (PIN) protein family of auxin transporters. Genome Biology. 2009;**10**:249

[10] Byrne ME, Barley R, Curtis M, Arroyo JM, Dunham M, Hudson A, Martienssen RA. Asymmetric leaves1 mediates leaf patterning and stem cell function in Arabidopsis. Nature. 2000;**408**:967-971

[11] Guo M, Thomas J, Collins G, Timmermans MCP. Direct repression of KNOX loci by the ASYMMETRIC LEAVES1 complex of Arabidopsis. The Plant Cell. 2008;**20**:48-58

[12] Li Z, Li B, Shen WH, Huang H, Dong A. TCP transcription factors interact with AS2 in the repression of class-I KNOX genes in Arabidopsis thaliana. The Plant Journal. 2012;**71**: 99-107

[13] Hay A, Barkoulas M, Tsiantis M. ASYMMETRIC LEAVES1 and auxin activities converge to repress BREVIPEDICELLUS expression and promote leaf development in Arabidopsis. Development. 2006;**133**:3955-3961

[14] Tsiantis M, Hay A. Comparative plant development: the time of the leaf? Nature Reviews Genetics. 2003;**4**:169-180

[15] Hareven D, Gutfinger T, Parnis A, Eshed Y, Lifschitz E. The making of a compound leaf: Genetic manipulation of leaf architecture in tomato. Cell. 1996;**84**:735-744

[16] Shani E, Burko Y, Ben-Yaakov L, Berger Y, Amsellem Z, Goldshmidt A, Sharon E, Ori N. Stage-specific regulation of Solanum lycopersicum leaf maturation by class 1 KNOTTED1-LIKE HOMEOBOX proteins. The Plant Cell. 2009;**21**:3078-3092

[17] Scarpella E, Barkoulas M, Tsiantis M. Control of leaf and vein development by auxin. Cold Spring Harbor Perspectives in Biology. 2010;**2**:a001511

[18] Yamaguchi H, Watanabe M, Sato S, Kanbayashi Y. Yielding ability of erect- and narrow-leaved rice mutant in heavy manuring and dense planting culture. Radioisotopes. 1979;**28**:734-738

[19] Chen M, Luo J, Shao G, Wei X, Tang S, Sheng Z, Song J, Hu P. Fine mapping of a major QTL for flag leaf width in rice, qFLW4, which might be caused by alternative splicing of NAL1. Plant Cell Reports. 2012;**31**:863-872

[20] Fujita D, Trijatmiko KR, Tagle AG, Sapasap MV, Koide Y, Sasaki K, Tsakirpaloglou N, Gannaban RB, Nishimura T, Yanagihara S, Fukuta Y, Koshiba T, Slamet-Loedin IH, Ishimaru T, Kobayashi N. NAL1 allele from a rice landrace greatly increases yield in modern indica cultivars. Proceedings of the National Academy of Sciences. 2013;**110**:20431-20436

[21] Takai T, Adachi S, Taguchi-Shiobara F, Sanoh-Arai Y, Iwasawa N, Yoshinaga S, Hirose S, Taniguchi Y, Yamanouchi U, Wu J, Matsumoto T, Sugimoto K, Kondo K, Ikka T, Ando T, Kono I, Ito S, Shomura A, Ookawa T, Hirasawa T, Yano M, Kondo M, Yamamoto T. A natural variant of NAL1, selected in high-yield rice breeding programs, pleiotropically increases photosynthesis rate. Scientific Reports. 2013;**3**:2149

[22] Zhang GH, Li SY, Wang L, Ye WJ, Zeng DL, Rao YC, Peng YL, Hu J, Yang YL, Xu, J, Ren DY, Gao ZY, Zhu L, Dong GJ, Hu XM, Yan MX, Guo LB, Li CY, Qian Q. LSCHL4 from Japonica Cultivar, which is allelic to NAL1, increases yield of indica super rice 93-11. Molecular Plant. 2014;**7**:1350-1364

[23] Taguchi-Shiobara F, Ota T, Ebana K, Ookawa T, Yamasaki M, Tanabata T, Yamanouchi U, Wu J, Ono N, Nonoue Y, Nagata K, Fukuoka S, Hirabayashi H, Yamamoto T, Yano M. Natural variation in the flag leaf morphology of rice due to a mutation of the NARROW LEAF 1 gene in Oryza sativa L. genetics. 2015;**201**:795-808

[24] Ljung K, Bhalerao RP, Sandberg G. Sites and homeostatic control of auxin biosynthesis in Arabidopsis during vegetative growth. The Plant Journal. 2001;**28**:465-474

[25] Benjamins R, Scheres B. Auxin: The looping star in plant development. Annu. Rev. Plant Biol. 2008;**59**:443-465

[26] Finet C, Jaillais Y. AUXOLOGY: When auxin meets plant evo-devo. Developmental Biology. 2012;**369**:19-31

[27] Zažímalová E, Murphy A, Yang H, Hoyerová K, Hošek P. Auxin transporters: why so many?. Cold Spring Harbor Perspectives in Biology. 2010;**2**:a001552

[28] Leyser O. Auxin, self-organisation, and the colonial nature of plants. Current Biology. 2011;**21**:331-337

[29] Vernoux T, Kronenberger J, Grandjean O, Laufs P, Traas J. PIN-FORMED 1 regulates cell fate at the periphery of the shoot apical meristem. Development. 2000;**127**:5157-5165

[30] Bainbridge K, Guyomarc'h S, Bayer E, Swarup R, Bennett M, Mandel T, Kuhlemeier C. Auxin influx carriers stabilize phyllotactic patterning. Genes and Development. 2008; **22**:810-823

[31] Scarpella E, Marcos D, Friml J, Berleth T. Control of leaf vascular patterning by polar auxin transport. Genes & Development. 2006;**20**:1015-1027

[32] Kitomi Y, Ogawa A, Kitano H, Inukai Y. CRL4 regulates crown root formation through auxin transport in rice. Plant Root. 2008;**2**:19-28

[33] Möller B, Weijers D. Auxin control of embryo patterning. Cold Spring Harbor Perspectives in Biology. 2009;**1**:a001545

[34] Forestan C, Meda S, Varotto S. ZmPIN1-mediated auxin transport is related to cellular differentiation during maize embryogenesis and endosperm development. Plant Physiology. 2010;**152**:1373-1390

[35] Mei Y, Jia WJ, Chu YJ, Xue HW. Arabidopsis phosphatidylinositol monophosphate 5-kinase 2 is involved in root gravitropism through regulation of polar auxin transport by affecting the cycling of PIN proteins. Cell Research. 2012;**22**:581-597

[36] Prusinkiewicz P, Crawford S, Smith RS, Ljung K, Bennett T, Ongaro V, Leyser O. Control of bud activation by an auxin transport switch. Proceedings of the National Academy of Sciences. 2009;**106**:17431-17436

[37] Vanneste S, Friml J. Auxin: A trigger for change in plant development. Cell. 2009;**136**:1005-1016

[38] Sugawara S, Hishiyama S, Jikumaru Y, Hanada A, Nishimura T, Koshiba T, Zhao Y, Kamiya Y, Kasahara H. Biochemical analyses of indole-3-acetaldoxime-dependent auxin biosynthesis in Arabidopsis. Proceedings of the National Academy of Sciences. 2009;**106**:5430-5435

[39] Zhao Y. Auxin biosynthesis and its role in plant development. Annual Review of Plant Biology. 2010;**61**:49-64

[40] Mashiguchi K, Tanaka K, Sakai T, Sugawara S, Kawaide H, Natsume M, Hanada A, Yaeno T, Shirasu K, Yao H, McSteen P, Zhao Y, Hayashi K, Kamiya Y, Kasahara H. The main auxin biosynthesis pathway in Arabidopsis. Proceedings of the National Academy of Sciences. 2011;**108**:18512-18517

[41] Zhao Y, Christensen SK, Fankhauser C, Cashman JR, Cohen JD, Weigel D, Chory J. A role for flavin monooxygenase-like enzymes in auxin biosynthesis. Science. 2001;**291**:306-309

[42] Won C, Shen X, Mashiguchi K, Zheng Z, Dai X, Cheng Y, Kasahara H, Kamiya Y, Chory J, Zhao Y. Conversion of tryptophan to indole-3-acetic acid by TRYPTOPHAN AMINOTRANSFERASES OF ARABIDOPSIS and YUCCAs in Arabidopsis. Proceedings of the National Academy of Sciences. 2011;**108**:18518-18523

[43] Stepanova AN, Robertson-Hoyt J, Yun J, Benavente LM, Xie DY, Doležal K, Schlereth A, Jürgens G, Alonso JM. TAA1-mediated auxin biosynthesis is essential for hormone crosstalk and plant development. Cell. 2008;**133**:177-191

[44] Stepanova AN, Yun J, Robles LM, Novak O, He W, Guo H, Ljung K, Alonso JM. The Arabidopsis YUCCA1 flavin monooxygenase functions in the indole-3-pyruvic acid branch of auxin biosynthesis. The Plant Cell. 2011;**23**:3961-3973

[45] Dai X, Mashiguchi K, Chen Q, Kasahara H, Kamiya Y, Ojha S, DuBois J, Ballou D, Zhao Y. The biochemical mechanism of auxin biosynthesis by an Arabidopsis YUCCA flavin-containing monooxygenase. Journal of Biological Chemistry. 2013;**288**:1448-1457

[46] Cheng Y, Dai X, Zhao Y. Auxin biosynthesis by the YUCCA flavin monooxygenases controls the formation of floral organs and vascular tissues in Arabidopsis. Genes and Development. 2006;**20**:1790-1799

[47] Gallavotti A, Barazesh S, Malcomber S, Hall D, Jackson D, Schmidt RJ, McSteen P. sparse inflorescence 1 encodes a monocot-specific YUCCA-like gene required for vegetative and reproductive development in maize. Proceedings of the National Academy of Sciences. 2008;**105**:15196-15201

[48] Phillips KA, Skirpan AL, Liu X, Christensen A, Slewinski TL, Hudson C, Barazesh S, Cohen JD, Malcomber S, McSteen P. vanishing tassel 2 encodes a grass-specific Tryptophan Aminotransferase required for vegetative and reproductive development in maize. The Plant Cell. 2011;**23**:550-566

[49] Fujino K, Matsuda Y, Ozawa K, Takeshi N, Koshiba T, Fraaije MW, Sekiguchi H. NARROW LEAF 7 controls leaf shape mediated by auxin in rice. Molecular Genetics and Genomics. 2008;**279**:499-507

[50] Yoshikawa T, Ito M, Sumikura T, Nakayama A, Nishimura T, Kitano H, Yamaguchi I, Koshiba T, Hibara K, Nagato Y, Itoh J. The rice FISH BONE gene encodes a tryptophan aminotransferase, which affects pleiotropic auxin-related processes. The Plant Journal. 2014;**78**:927-936

[51] Sazuka T, Kamiya N, Nishimura T, Ohmae K, Sato Y, Imamura K, Nagato Y, Koshiba T, Nagamura Y, Ashikari M, Kitano H, Matsuoka M. A rice tryptophan deficient dwarf mutant, tdd1, contains a reduced level of indole acetic acid and develops abnormal flowers and organless embryos. The Plant Journal. 2009;**60**:227-241

[52] Woo YM, Park HJ, Su'udi M, Yang JI, Park JJ, Back K, Park YM, An G. Constitutively wilted 1, a member of the rice YUCCA gene family, is required for maintaining water homeostasis and an appropriate root to shoot ratio. Plant Molecular Biology. 2007;**65**:125-136

[53] Abu-Zaitoon YM, Bennett K, Normanly J, Nonhebel HM. A large increase in IAA during development of rice grains correlates with the expression of tryptophan aminotransferase OsTAR1 and a grain-specific YUCCA. Physiologia Plantarum. 2012;**146**:487-499

[54] Qi J, Qian Q, Bu Q, Li S, Chen Q, Sun J, Liang W, Zhou Y, Chu C, Li X, Ren F, Palme K, Zhao B, Chen J, Chen M, Li C. Mutation of the rice Narrow leaf1 gene, which encodes a novel protein, affects vein patterning and polar auxin transport. Plant Physiology. 2008;**147**:1947-1959

[55] Jiang D, Fang J, Lou L, Zhao J, Yuan S, Yin L, Sun W, Peng L, Guo B, Li X. Characterization of a null allelic mutant of the rice NAL1 gene reveals its role in regulating cell division. PLoS ONE. 2015;**10**. DOI: 10.1371/journal.pone.0118169

[56] Scheller HV, Ulvskov P. Hemicelluloses. Annual Review of Plant Biology. 2010;**61**:263-289

[57] Richmond T. Higher plant cellulose synthases. Genome Biology. 2000;**1**:reviews3001.1-1.6.

[58] Kurek I, Kawagoe Y, Jacob-Wilk D, Doblin M, Delmer D. Dimerization of cotton fiber cellulose synthase catalytic subunits occurs via oxidation of the zinc-binding domains. Proceedings of the National Academy of Sciences. 2002;**99**:11109-11114

[59] Taylor NG. Cellulose biosynthesis and deposition in higher plants. New Phytologist. 2008;**178**:239-252

[60] Richmond TA, Somerville CR. The cellulose synthase superfamily. Plant Physiology. 2000;**124**:495-498

[61] Hazen SP, Scott-Craig JS, Walton JD. Cellulose synthase-like genes of rice. Plant Physiology. 2002;**128**:336-340

[62] Fincher GB. Revolutionary times in our understanding of cell wall biosynthesis and remodeling in the grasses. Plant Physiology. 2009;**149**:27-37

[63] Dhugga KS, Barreiro R, Whitten B, Stecca K, Hazebroek J, Randhawa GS, Dolan M, Kinney AJ, Tomes D, Nichols S, Anderson P. Guar seed beta-mannan synthase is a member of the cellulose synthase super gene family. Science. 2004;**303**:363-366

[64] Liepman AH, Wilkerson CG, Keegstra K. Expression of cellulose synthase-like (Csl) genes in insect cells reveals that CslA family members encode mannan synthases. Proceedings of the National Academy of Sciences. 2005;**102**:2221-2226

[65] Suzuki S, Li L, Sun YH, Chiang VL. The cellulose synthase gene superfamily and biochemical functions of xylem-specific cellulose synthase-like genes in Populus trichocarpa. Plant Physiology. 2006;**142**:1233-1245

[66] Burton RA, Wilson SM, Hrmova M, Harvey AJ, Shirley NJ, Medhurst A, Stone BA, Newbigin EJ, Bacic A, Fincher GB. Cellulose synthase-like CslF genes mediate the synthesis of cell wall (1,3;1,4)-beta-D-glucans. Science. 2006;**311**:1940-1942

[67] Doblin MS, Pettolino FA, Wilson SM, Campbell R, Burton RA, Fincher GB, Newbigin E, Bacic A. A barley cellulose synthase-like CSLH gene mediates (1,3;1,4)-beta-D-glucan synthesis in transgenic Arabidopsis. Proceedings of the National Academy of Sciences. 2009;**106**:5996-6001

[68] Cocuron JC, Lerouxel O, Drakakaki G, Alonso AP, Liepman AH, Keegstra K, Raikhel N, Wilkerson CG. A gene from the cellulose synthase-like C family encodes a β-1,4-glucan synthase. Proceedings of the National Academy of Sciences. 2007;**104**:8550-8555

[69] Dwivany FM, Yulia D, Burton RA, Shirley NJ, Wilson SM, Fincher GB, Bacic A, Newbigin E, Doblin MS. The Cellulose-synthase like C (CslC) family of barley includes members that are integral membrane proteins targeted to the plasma membrane. Molecular Plant. 2009;**2**:1025-1039

[70] Roberts AW, Bushoven JT. The cellulose synthase (CESA) gene superfamily of the moss Physcomitrella patens. Plant Molecular Biology. 2007;**63**:207-219

[71] Hunter CT, Kirienko DH, Sylvester AW, Peter GF, McCarty DR, Koch KE. Cellulose synthase-like D1 is integral to normal cell division, expansion, and leaf development in maize. Plant Physiology. 2012;**158**:708-724

[72] Yoshikawa T, Eiguchi M, Hibara K, Ito J, Nagato Y. Rice SLENDER LEAF1 gene encodes cellulose synthase-like D4 and is specifically expressed in M-phase cells to regulate cell proliferation. Journal of Experimental Botany. 2013;**64**:2049-2061

[73] Bernal AJ, Yoo CM, Mutwil M, Jensen JK, Hou G, Blaukopf C, Sørensen I, Blancaflor EB, Scheller HV, Willats WGT. Functional analysis of the cellulose synthase-like genes Csld1, Csld2, and Csld4 in tip-growing Arabidopsis cells. Plant Physiology. 2008;**148**:1238-1253

[74] Favery B, Ryan E, Foreman J, Linstead P, Boundonck K, Steer M, Shaw P, Dolan L. KOJAK encodes a cellulose synthase-like protein required for root hair cell morphogenesis in Arabidopsis. Genes and Development. 2001;**15**:79-89

[75] Wang X, Cnops G, Vanderhaeghen R, De Block S, Van Montagu M, Van Lijsebettens M. AtCSLD3, a cellulose synthase-like gene important for root hair growth in Arabidopsis. Plant Physiology. 2001;**126**:575-586

[76] Kim CM, Park SH, Je BI, Park SH, Park SJ, Piao HL, Eun MY, Dolan L, Han C. OsCsld1, a cellulose synthase-like D1 gene, is required for root hair morphogenesis in rice. Plant Physiology. 2007;**143**:1220-1230

[77] Penning BW, Hunter CTIII, Tyengwa R, Eveland AL, Dugard CK, Olek AT, Vermerris W, Koch KE, McCarty DR, Davis MF, Thomas SR, McCann MC, Carpita NC. Genetic resources for maize cell wall biology. Plant Physiology. 2009;**151**:1703-1728

[78] Li M, Xiong G, Li R, Cui J, Tang D, Zhang B, Pauly M, Cheng Z, Zhou Y. Rice cellulose synthase-like D4 is essential for normal cell-wall biosynthesis and plant growth. The Plant Journal. 2009;**60**:1055-1069

[79] Hu J, Zhu L, Zeng D, Gao Z, Guo L, Fang Y, Zhang G, Dong G, Yan M, Liu J, Qian Q. Identification and characterization of NARROW AND ROLLED LEAF 1, a novel gene regulating leaf morphology and plant architecture in rice. Plant Molecular Biology. 2010;**73**:283-292

[80] Luan W, Liu Y, Zhang F, Song Y, Wang Z, Peng Y, Sun Z. OsCD1 encodes a putative member of the cellulose synthase-like D sub-family and is essential for rice plant architecture and growth. Plant Biotechnology Journal. 2010;**9**:513-524

[81] Ding Z, Lin Z, Li Q, Wu H, Xiang C, Wang J. DNL1, encodes cellulose synthase-like D4, is a major QTL for plant height and leaf width in rice (Oryza sativa L.). Biochemical and Biophysical Research Communications. 2015;**457**:133-140

[82] Shi L, Wei XJ, Adedze YM, Sheng ZH, Tang SQ, Hu PS, Wang JL. Characterization and gene cloning of the rice (Oryza sativa L.) dwarf and narrow-leaf mutant dnl3. Genetics and Molecular Research. 2016;**15**. DOI: 10.4238/gmr.15038731

[83] Horiguchi G, Tsukaya H. Organ size regulation in plants: insights from compensation. Frontiers in Plant Science. 2011;**2**:1-6

[84] Lukowitz W, Mayer U, Jügens G. Cytokinesis in the Arabidopsis embryo involves the synthaxin-related KNOLLE gene proruct. Cell. 1996;**84**:61-71

[85] Zuo J, Niu QW, Nishizawa N, Wu Y, Kost B, Chua NH. KORRIGAN, an Arabidopsis endo-1,4-β-glucanase, localizes to the cell plate by polarized ta-getting and is essential for cytokinesis. The Plant Cell. 2000;**12**:1137-1152

[86] Strompen G, Kasmi FE, Richter S, Lukowitz W, Assaad FF, Jügens G, Mayer U. The Arabidopsis HINKEL gene encodes a kinesin-related protein involved in cytokinesis and is expressed in a cell cycle-dependent manner. Current Biology. 2002;**12**:153-158

[87] Yin L, Verhertbruggen Y, Oikawa A, Manisseri C, Knierim B, Prak L, Jensen JK, Knox JP, Auer M, Willats WGT, Scheller HV. The cooperative activities of CSLD2, CSLD3, and CSLD5 are required for normal Arabidopsis development. Molecular Plant. 2011;**4**:1024-1037

[88] Goubet F, Barton CJ, Mortimer JC, Yu X, Zhang Z, Miles GP, Richens J, Liepman AH, Seffen K, Dupree P. Cell wall glucomannan in Arabidopsis is synthesised by CSLA glycosyltransferases, and influences the progression of embryogenesis. Plant J. 2009;**60**:527-538

[89] Itoh J, Nonomura K, Ikeda K, Yamaki S, Inukai Y, Yamagishi H, Kitano H, Nagato Y. Rice plant development: from zygote to spikelet. Plant Cell Physiol. 2005;**46**:23-47

[90] Nakata M, Okada K. The leaf adaxial-abaxial boundary and lamina growth. Plants. 2013;**2**:174-202

[91] Heisler MG, Ohno C, Das P, Sieber P, Reddy GV, Long JA, Meyerowitz EM. Patterns of auxin transport and gene expression during primordium development revealed by live imaging of the Arabidopsis inflorescence meristem. Curr. Biol. 2005;**15**:1899-1911

[92] Vernoux T, Besnard F, Traas J. Auxin at the shoot apical meristem. Cold Spring Harbor Perspectives in Biology. 2010;**2**:a001487

[93] Yan S, Yan CJ, Zeng XH, Yang YC, Fang YW, Tian CY, Sun YW, Cheng ZK, Gu MH. ROLLED LEAF 9, encoding a GARP protein, regulates the leaf abaxial cell fate in rice. Plant Molecular Biology. 2008;**68**:239-250

[94] Zhang GH, Xu Q, Zhu XD, Qian Q, Xue HW. SHALLOT-LIKE1 is a KANADI transcription factor that modulates rice leaf rolling by regulating leaf abaxial cell development. Plant Cell. 2009;**21**:719-735

[95] Juarez MT, Kui JS, Thomas J, Heller BA, Timmermans MC. microRNA-mediated repression of rolled leaf1 specifies maize leaf polarity. Nature. 2004;**428**:84-88

[96] Freeling M. A conceptual framework for maize leaf development. Journal of Developmental Biology. 1992;**153**:44-58

[97] Sinha N, Hake S. The Knotted leaf blade is a mosaic of blade, sheath, and auricle identities. Developmental Genetics Journal. 1994;**15**:401-414

[98] Vollbrecht E, Veit B, Sinha N, Hake S. The developmental gene Knotted-1 is a member of a maize homeobox gene family. Nature. 1991;**350**:241-243

[99] Smith L, Greene B, Veit B, Hake S. A dominant mutation in the maize homeobox gene, knotted-1, causes its ectopic expression in leaf cells with altered fates. Development. 1992;**116**:21-30

[100] Jackson D, Veit B, Hake S. Expression of maize KNOTTED1 related homeobox genes in the shoot apical meristem predicts patterns of morphogenesis in the vegetative shoot. Development. 1994;**120**:405-413

[101] Foster T, Veit B, Hake S. Mosaic analysis of the dominant mutant, Gnarley1-R, reveals distinct lateral and transverse signaling pathways during maize leaf development. Development. 1999;**126**:305-313

[102] Ramirez J, Bolduc N, Lisch D, Hake S. Distal expression of knotted1 in maize leaves leads to reestablishment of proximal/distal patterning and leaf dissection. Plant Physiology. 2009;**151**:1878-1888

[103] Müller KJ, Romano N, Gerstner O, Garcia-Maroto F, Pozzi C, Salamini F, Rohde W. The barley Hooded mutation caused by a duplication in a homeobox gene intron. Nature. 1995;**374**:727-730

[104] Schneeberger RG, Becraft PW, Hake S, Freeling M. Ectopic expression of the knox homeo box gene rough sheath1 alters cell fate in the maize leaf. Genes & Development. 1995;**9**:2292-2304

[105] Ha CM, Kim GT, Kim BC, Jun JH, Soh MS, Ueno Y, Machida Y, Tsukaya H, Nam HG. The BLADE-ON-PETIOLE 1 gene controls leaf pattern formation through the modulation of meristematic activity in Arabidopsis. Development. 2003;**130**:161-172

[106] Bayer EM, Smith RS, Mandel T, Nakayama N, Sauer M, Prusinkiewicz P, Kuhlemeier C. Integration of transportbased models for phyllotaxis and midvein formation. Genes & Development. 2009;**23**:373-384

[107] Becraft PW, Bongard-Pierce DK, Sylvester AW, Poethig RS, Freeling M. The liguleless-1 gene acts tissue specifically in maize leaf development. Developmental Biology. 1990;**141**:220-232

[108] Becraft PW, Freeling M. Sectors of liguleless-1 tissue interrupt an inductive signal during maize leaf development. Plant Cell. 1991;**3**:801-807

[109] Harper L, Freeling M. Interactions of liguleless1 and liguleless2 function during ligule induction in maize. Genetics. 1996;**144**:1871-1882

[110] Moreno MA, Harper LC, Krueger RW, Dellaporta SL, Freeling M. liguleless1 encodes a nuclear-localized protein required for induction of ligules and auricles during maize leaf organogenesis. Genes & Development. 1997;**11**:616-628

[111] Walsh J, Water CA, Freeling M. The maize gene liguleless2 encodes a basic leucine zipper protein involved in the establishment of the leaf blade-sheath boundary. Genes & Development. 1998;**12**:208-218

[112] Foster T, Hay A, Johnston R, Hake S. The establishment of axial patterning in the maize leaf. Development. 2004;**131**:3921-3929

[113] Johnston R, Wang M, Sun Q, Sylvester AW, Hake S, Scanlon MJ. Transcriptomic analyses indicate that maize ligule development recapitulates gene expression patterns that occur during lateral organ initiation. Plant Cell. 2014;**26**:4718-4732

[114] Muehlbauer GJ, Fowler JE, Girard L, Tyers R, Harper L, Freeling M. Ectopic expression of the maize homeobox gene liguleless3 alters cell fates in the leaf. Plant Physiol. 1999;**119**:651-662

[115] Bauer P, Lubkowitz M, Tyers R, Nemoto K, Meeley RB, Goff SA, Freeling M. Regulation and a conserved intron sequence of liguleless3/4 knox class-I homeobox genes in grasses. Planta. 2004;**219**:359-368

[116] Yamaguchi T, Nagasawa N, Kawasaki S, Matsuoka M, Nagato Y, Hirano HY. The YABBY gene DROOPING LEAF regulates carpel specification and midrib development in Oryza sativa. Plant Cell. 2004;**16**:500-509

[117] Nardmann J, Ji J, Werr W, Scanlon MJ. The maize duplicated genes narrow sheath1 and narrow sheath2 encodes a conserved homeobox gene function in a lateral domain of shoot apical meristems. Development. 2004;**131**:2827-2839

[118] Scanlon MJ, Freeling M. The narrow sheath leaf domain: a genetic tool used to reveal developmental homologies among modified maize organs. The Plant Journal. 1998;**13**:547-561

[119] Scanlon MJ, Schneeberger RG, Freeling M. The maize mutant narrow sheath fails to establish leaf margin identity in a meristematic domain. Development. 1996;**122**:1683-1691

[120] Scanlon MJ. NARROW SHEATH1 functions from two meristematic foci during founder-cell recruitment in maize leaf development. Development. 2000;**127**:4573-4585

[121] Scanlon MJ, Freeling M. Clonal sectors reveal that a specific meristematic domain is not utilized in the maize mutant narrow sheath. Developmental Biology. 1997;**182**:52-66

[122] Scanlon MJ, Chen KD, McKnight IV CC. The narrow sheath duplicate gene: sectors of dual aneuploidy reveal ancestrally conserved gene functions during maize leaf development. Genetics. 2000;**155**:1379-1389

[123] Yoshikawa T, Tanaka SY, Masumoto Y, Nobori N, Ishii H, Hibara K, Itoh J, Tanisaka T, Taketa S. Barley NARROW LEAFED DWARF1 encoding a WUSCHEL-RELATED HOMEOBOX 3 (WOX3) regulates the marginal development of lateral organs. Breeding Science. 2016; **66**:416-424

[124] The Rice Chromosomes 11 and 12 Sequencing Consortia. The sequence of rice chromosomes 11 and 12, rich in disease resistance genes and recent gene duplications. BMC Biology. 2005;**3**:20

[125] Cho SH, Yoo SC, Zhang H, Pandeya D, Koh HJ, Hwang JY, Kim GT, Paek NC. The rice narrow leaf2 and narrow leaf3 loci encode WUSCHEL-related homeobox 3A (OsWOX3A) and function in leaf, spikelet, tiller and lateral root development. New Phytologist. 2013;**198**:1071-1084

[126] Ishiwata A, Ozawa M, Nagasaki H, Kato M, Noda Y, Yamaguchi T, Nosaka M, Shimizu-Sato S, Nagasaki A, Maekawa M, Hirano HY, Sato Y. Two WUSCHEL-related homeobox genes, narrow leaf2 and narrow leaf3, control leaf width in rice. Plant Cell Physiol. 2013;**54**:779-792

[127] Troll W. Concerning the morphological significance of the so-called vorlaeuferspitze of monocot leaves. A contribution to the typology of monocot leaves. Beitr. Biol. Pflanz. 1955;**31**:525-558

[128] Haecker A, Gross-Hardt R, Geiges B, Sarkar A, Breuninger H, Herrmann M, Laux T. Expression dynamics of WOX genes mark cell fate decisions during early embryonic patterning in Arabidopsis thaliana. Development. 2004;**131**:657-668

[129] Vandenbussche M, Horstman A, Zethof J, Koes R, Rijpkema AS, Gerats T. Differential recruitment of WOX transcription factors for lateral development and organ fusion in Petunia and Arabidopsis. Plant Cell. 2009;**21**:2269-2283

[130] Nakata M, Matsumoto N, Tsugeki R, Rikirsch E, Laux T, Okada K. Roles of the middle domain-specific WUSCHEL-RELATED HOMEOBOX genes in early development of leaves in Arabidopsis. Plant Cell. 2012;**24**:519-535

[131] Schneeberger R, Tsiantis M, Freeling M, Langdale JA. The rough sheath2 gene negatively regulates homeobox gene expression during maize leaf development. Development. 1998;**125**:2857-2865

[132] Moon J, Candela H, Hake S. The Liguleless narrow mutation affects proximal-distal signaling and leaf growth. Development. 2013;**140**:405-412

[133] Hay A, Hake S. The dominant mutant Wavy auricle in blade1 disrupts patterning in a lateral domain of the maize leaf. Plant Physiol. 2004;**135**:300-308

[134] Lewis MW, Bolduc N, Hake K, Htike Y, Hay A, Candela H, Hake S. Gene regulatory interactions at lateral organ boundaries in maize. Development. 2014;**141**:4590-4597

[135] Liu X, Li M, Liu K, Tang D, Sun M, Li Y, Shen Y, Du G, Cheng Z. Semi-Rolled Leaf 2 modulates rice leaf rolling by regulating abaxial side cell differentiation. Journal of Experimental Botany. 2016;**67**:2139-2150

[136] Li YY, Shen A, Xiong W, Sun QL, Luo Q, Song T, Li ZL, Luan WJ. Overexpression of OsHox32 results in pleiotropic effects on plant type architecture and leaf development in rice. Rice. 2016;**9**:46. DOI: 10.1186/s12284-016-0118-1

[137] Li C, Zou X, Zhang C, Shao Q, Liu J, Liu B, Li H, Zhao T. OsLBD3-7 Overexpression induced adaxially rolled leaves in rice. PLoS One. 2016;**11**:e0156413. DOI: 10.1371/journal.pone.0156413

[138] Henderson DC, Muehlbauer GJ, Scanlon MJ. Radial leaves of the maize mutant ragged seedling2 retain dorsiventral anatomy. Developmental Biology. 2005;**282**:455-466

[139] Itoh J, Sato Y, Nagato Y. The SHOOT ORGANIZATION2 gene coordinates leaf domain development along the central-marginal axis in rice. Plant Cell Physiol. 2008;**49**: 1226-1236

[140] Timmermans MC, Schultes NP, Jankovsky JP, Nelson T. Leafbladeless1 is required for dorsoventrality of lateral organs in maize. Development. 1998;**125**:2813-2823

[141] Nogueira FT, Madi S, Chitwood DH, Juarez MT, Timmermans MC. Two small regulatory RNAs establish opposing fates of a developmental axis. Genes & Development. 2007;**21**:750-755

[142] Satoh N, Itoh J, Nagato Y. The SHOOTLESS2 and SHOOTLESS1 genes are involved in both initiation and maintenance of the shoot apical meristem through regulating the number of indeterminate cells. Genetics. 2003;**164**:335-346

[143] Nagasaki H, Itoh J, Hayashi K, Hibara K, Satoh-Nagasawa N, Nosaka M, Mukouhata M, Ashikari M, Kitano H, Matsuoka M, Nagato Y, Sato Y. The small interfering RNA production pathway is required for shoot meristem initiation in rice. Proceedings of the National Academy of Sciences. 2007;**104**:14867-14871

[144] Itoh JI, Kitano H, Matsuoka M, Nagato Y. Shoot organization genes regulate shoot apical meristem organization and the pattern of leaf primordium initiation in rice. Plant Cell. 2000;**12**:2161-2174

[145] Sarojam R, Sappl PG, Goldshmidt A, Efroni I, Floyd SK, Eshed Y, Bowman JK. Differentiating Arabidopsis shoots from leaves by combined YABBY activities. Plant Cell. 2010;**22**:2113-2130

[146] Stahle MI, Kuehlich J, Staron L, von Arnim AG, Golz JF. YABBYs and the transcriptional corepressors LEUNIG and LEUNIG_HOMOLOG maintain leaf polarity and meristem activity in Arabidopsis. Plant Cell. 2009;**21**:3105-3118

[147] Wang W, Xu B, Wang H, Li J, Huang H, Xu L. YUCCA genes are expressed in response to leaf adaxial-abaxial juxtaposition and are required for leaf margin development. Plant Physiol. 2011;**157**:1805-1819

[148] Kumar S, Stecher G, Tamura K. MEGA7: Molecular evolutionary genetics analysis version 7.0 for bigger datasets. Molecular Biology and Evolution. 2016;**33**:1870-1874

Chapter 2

Oxidative Enzyme Effects in Malt for Brewing

Makoto Kanauchi

Additional information is available at the end of the chapter

http://dx.doi.org/10.5772/intechopen.69803

Abstract

Malted barley is an important beer-brewing material that strongly affects brewing processes, the aroma, and the taste of beer. In addition to imparting a good aroma, malt not only generates substrates and enzymes, such as starches and some amylase, for alcohol production but also generates beer-quality-degrading substances and enzymes. Four oxidases are specifically addressed in this chapter. First, thiol oxidase in malt is described. The activity of thiol oxidase decreases during malt storage. Next, ascorbate peroxidase was investigated. It has been detected in the acrospires and aleurones of germinating barley. The enzyme has extremely high affinity for hydrogen peroxide. Also, ascorbic acid oxidase (AAO) was investigated. It is developed in the embryo tissues of barley during steeping and during the initial stages of germination. The addition of ascorbic acid to mash leads to the survival of higher levels of polyphenol and thiols into wort and a reduced color in that wort. Finally, oxalate oxidase in barley kernels is described. It is probably less important than other oxidases in scavenging oxygen from mashes, because the enzyme has low affinity for oxygen. Beer quality is expected to be improved by the regulation of oxidant enzymes, such as thiol oxidase or AAO, oxalate oxidase, or substrates, such as oxygen.

Keywords: malt, oxidation enzymes, thiol, ascorbic acid, hydrogen peroxide, oxalic acid

1. Introduction

1.1. Malt science

Malted barley is an important material used in producing beer. Maltose is made from starch by the enzyme developed during barley germination. Yeast ferments sugars to alcohol. Components and enzymes of malt are important for beer quality. This chapter presents an explanation of cell wall oxidase in malt and germinated barley.

Kernels of barley (*Hordeum vulgare*) are of two types: two-row and six-row. Two-row barley is cultivated in Australia, Canada, and Europe and six-row barley is cultivated in the USA. The barley grain covers a glume. The endosperm is protected by the husk, which is fractured by a mechanical shock during malting. Commercial barleys have the husk removed from the grain. The husk is largely insoluble, forming a filter bed for lautering in the brewing process. The pericarp is waxy and waterproof. The testa acts as a permeable membrane. Water readily penetrates into the embryo through the testa during steeping. The seed of the barley and the embryo and aleurone layer are living. After the kernel absorbs water, gibberellin is excreted as a phytohormone from the germ. Furthermore, the signal of gibberellin leads to the production of enzymes, such as amylase, and to the synthesis of β-glucanase to decompose starch. The reaction transitions from endosperm verging of the aleurone or germ to internal endosperm for decomposition of starch [1].

Barley for brewing must meet six necessary conditions [1]:

1. Grains are large, with uniform size and shape.
2. Barley has a thin husk.
3. Grains contain uniform protein concentration and rich starch.
4. Grains have vigorous germination evenly.
5. Germinated grains have a high level of enzyme activity.
6. Germinated grains (malt) can saccharify easily and can ferment quickly for wort.

1.2. History of beer

Ancient beer called '*SIKARU*' was brewed from Emmer wheat in Mesopotamia in 3000 BC. In Egypt, bread made from germinated barley was dissolved with water and fermented in a vase. It is not modern beer. At about that time, '*GRUUT*,' a herbal mixture of *Myrica gale*, mugwort, or others, was used for beer production in Medieval Europe. Hops have been used for beer-brewing since the twelfth to the fifteenth century. Modern beer, with its fresh aroma, bitterness, and rich foaming is developed from the use of hops. The purpose of using hops has been the subject of various theories. However, hops might be useful mainly in preventing the growth of lactic acid bacteria [1].

1.3. Consumption of beer

Beer is the most consumed alcoholic beverage in the world. Annual global consumption was 146 million kL in 2015. People in China, who consumed 43 million kL of beer in 2015, consume more beer than is consumed in any other country. The second leading country for consumption is the USA, with 22 million kL of beer consumed, followed in order by Brazil, Germany, and Mexico. These five countries collectively consume more than 60% of all the beer produced. Japan is ranked seventh, with 5 million kL of beer consumption (Kirin Co. Ltd.) [2].

Japan has a unique system of liquor tax laws. Beer is categorized legally according to the ratio of malt and other starchy materials [1]. Beer is defined legally in Japan as described below:

(i) Beer is fermented from materials, such as malt, hops, and water.

(ii) Beer is fermented malt, hops, water, and other legally allowed materials, such as barley, rice, sorghum, corn, potato, starch, and sugars. However, the mass of the materials must be less than 50% of the mass of malt.

Before the law changed in 1996, legally allowed materials were less than one-third of the malt quantity. Japanese brewery companies sold low-malt, low-tax, and low-cost beer. After changing the law in 2006, beer or malt beverages came to be classified into three categories: first, the ratio of malt among all starch materials is greater than 50%; second, the ratio of malt among all starch materials is 25–50%; and third, the ratio of malt among all starch materials is less than 25%. Low-malt, low-tax, and low-cost beer is known as '*HAPPOUSHU*,' which is generally brewed from less than 25% malt among all starch materials [1].

2. Oxidative enzyme in malt

Generally, two types of oxidations occur in food: enzymic oxidation and nonenzymic oxidation. By enzymic oxidation, an apple changes color because of the effects of polyphenol oxidase. By nonenzymic oxidation, auto-oxidation of lipids occurs by the production of peroxide. Beer also has oxidants and two similar oxidation systems. Their oxidation types are important factors affecting beer quality. In this section, effects of oxidation enzymes in brewing processes are presented in **Table 1** [3–7].

2.1. Thiol oxidase

To avoid oxidation during germination, freshly kilned malt is typically not stored for more than 2–4 weeks. This storage inability presents problems, notably reduced rates of wort separation [8].

Flavor:
lipid oxidation; LOX or peroxidase [3]
Higher alcohol oxidation [3]
Forming haze:
Oxidation oxidized polyphenols tend to cross-link with protein [4]
Dark color:
Second source of coloring, oxidation of polyphenol or tannic material [4]
Difficult wort-run off:
Hordein protein-containing cysteine having sulpide bond. After Mash oxidizes during mashing, wort runs off a filter slowly, because of Hordein protein undergo oxidative polymerization as disulfide bond [5–7]

Table 1. Effects of oxidation enzymes on beer quality.

Freshly kilned malt is unsuitable for mashing [9]. Furthermore, saccharification also occurs less readily and the wort becomes turbid. It might also give rise to poor fermentation and haze in the beer. Therefore, storage for about three weeks before use is conducted as a precaution.

The oxidation of peptides having thiol-amino residue as cysteine produces gel proteins, which impede wort separation [7]. Reduced protein-containing thiol groups engenders acceleration of wort filtration [10]. Muller [11] reported the production of hydrogen peroxide in mashes attendant to the oxidation of thiol groups. Stephenson et al. [12] reported that some share of thiol oxidation occurring in mashes is enzyme-catalyzed. Malt extracts oxidized the cysteine to the equivalent dithiol cysteine. When malt extract was denatured by boiling, the interchange did not cause enzymic oxidation, such as thiol oxidase. The enzyme has Enzyme Commission number EC 1.8.3.2.

$$2\,RSH + O_2 \rightarrow RSSR + H_2O_2 \quad (1)$$

Bamforth et al. [13] reported the isolation of such an enzyme from malt and discussed some of its salient properties. Barley was germinated as described in an earlier report by Hoy et al. [14]. The green malt was dried at 60°C for 16 h. Thereafter, the malt was stored in incubators at 10, 20, or 30°C. Barley or germinated barley was ground in a coffee grinder. Then the crude extract was extracted by stirring on ice in buffer containing 0.25 M NaCl for 1 h. The slurry was then strained through cheesecloth. The resulting liquid was centrifuged at 13,000 rpm for 10 min. Denatured samples were the supernatants of crude extracts autoclaved using a laboratory autoclave switched to the "liquid" setting adjusted to sterilize for 15 min. Furthermore, thiol oxidation assay was conducted on extract of raw or malted barley that were incubated at 25°C in a potassium phosphate buffer (pH 7.5)-containing ethylenediaminetetraacetic acid (EDTA) together with cysteine or another reduced substrate (dithiothreitol, glutathione, or mercaptoethanol). Aliquots (0.15 mL) were removed at intervals for the determination of thiol using Ellman's [15] reagent (5,5′-dithiobis (2-nitrobenzoic acid) [DTNB]) by measuring the light absorbance in a spectrophotometer at 412 nm. To purify thiol oxidase, malt milled as described earlier was extracted for 3 h at 4°C with borate buffer (pH 9.0)-containing 2 mM EDTA and 0.5% NaCl. After centrifugation of the extract, thiol oxidase was purified by ammonium sulfate precipitation, a column of DEAE gel (Macro-Prep DEAE, 25 mm × 300 mm, Bio-Rad, Hercules, CA, USA), and a size-exclusion column (10 mm × 350 mm, P-100 gel, Bio-Rad).

The enzyme had a molecular weight of 35,700 Da. Thiol oxidase had a pH optimum of approximately pH 8.0. During mashing, the pH of the enzymic activity was low. The pH in mash, such as pH 5.0 was too low for the thiol oxidase reaction. Storage at 20°C led to the virtual elimination of activity after 4 weeks, but at 30°C, the activity was lost completely after 2 weeks. The enzyme decreased the rate of filtration of gel proteins because it promotes the cross-linking of these molecules.

The extent of the reduction of filtration was also greatest under alkaline conditions. However, it did not agree with the optimum pH as determined in the enzyme assay. The very low activity of enzyme extract at pH 5.0 is worth exploring a little more. Bearing in mind this possibility, the presumed enzyme catalyzing the oxidation of thiols was observed to be resistant to more modest degrees of heat. This enzyme has heat tolerance. The residual appearance enzyme activity at 100°C was approximately 50% of that of the nonheat enzyme. High thermal tolerance has been claimed for similar enzymes in other organisms [16–18].

In speculation of enzyme specificity, the enzyme was extremely active with cysteine as the substrate. It had much less activity with reduced glutathione (a tripeptide of glutamate, cysteine, and glycine), mercaptoethanol, and dithiothreitol (thiol compound often used in biochemical systems). Malt contains a heat-stable thiol oxidase. It is far from its optimum pH (pH 8.0) when at pH 5, which is its general mashing pH. It can be predicted that lowering the mashing pH would lessen the extent of its action. The activity of thiol oxidase lessens during malt storage, which suggests that this is part of the explanation for why stored malt displays better wort separation than newly kilned malt.

2.2. Ascorbate peroxidase

Kanauchi and Bamforth [19] examined ascorbate peroxidase in barley and germinated barley. The reaction of L-ascorbate peroxidase (EC 1.11.1.11; APX), an enzyme that exists in plants and microorganisms [20], is shown below.

$$\text{Ascorbate} + H_2O_2 \rightarrow \text{Dehydroascorbate} + H_2O \tag{2}$$

It plays a role in removing damaging reactive oxygen species. This enzyme in barley grain has attracted little study [21].

No attention of researchers has been devoted to the relative importance of this enzyme in malting and brewing. Kanauchi and Bamforth [19] reported the effects on malting and brewing that are exerted by peroxidase functions.

Preparation of malt barley included its germination, as described by Hoy et al. [14]. The green malt was dried by lyophilization or by kilning at 60°C for 48 h. Barley or malt was ground in a food processor and was extracted by stirring on ice in 2.5 volumes of phosphate buffer (pH 7.0)-containing NaCl and EDTA. The slurry was then strained through cheesecloth. The resulting liquid was centrifuged at 4000× *g* to prepare a crude enzyme solution. Barley or malt was steeped in a hydrogen peroxide/ascorbic acid solution in phosphate buffer at 30°C for 1–4 h. After reaction, residual peroxide was detected using ferricyanide/ferric chloride or using a 3,35′-diaminobenzidine solution-containing peroxidase [22]. APX was assayed according to a procedure described by Nakano and Asada [23]. Ascorbic acid was assayed using an F-kit (Roche Diagnostics Corp.).

The development of APX in germinating barley started steeping immediately. It is extensive between the second and third days of germination. Thereafter, the levels start to decline. The enzyme level was raised considerably by the accompanying increased level of ascorbic acid (**Figure 1**). The enzyme is located primarily in the acrospire and the starchy endosperm (**Figure 2A–C**). Actually, the enzyme is abundant in the starchy endosperm. The enzyme survives light kilning (**Figure 3**). The activity of ascorbate peroxidase in germinated barley was 239.4 (U/g dry malt). After kilning, it was 159.6 (U/g dry malt) after 16 h; after 48 h, it was 136.5 (U/g dry malt). The enzyme reduces to 57.0–66.7% (data not shown).

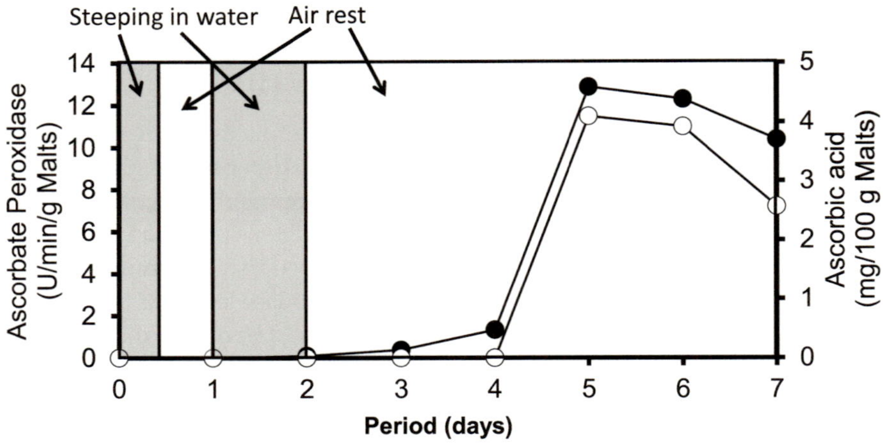

Figure 1. Levels of ascorbate peroxidase (•) and ascorbic acid (○) during the malting of barley.

To show the location of APX in kernels, endosperm was cut from sterile, dehusked barley at a distance of 2 mm. The slices were incubated at room temperature in sterilized gibberellic acid (10^{-5} M). After incubation, 5 mL of phosphate buffer (pH 7.0)-containing NaCl was added. Then the tissues were ground with a mortar and pestle. The extraction was prepared by centrifugation. A quantitative analysis of the distribution of the enzyme and also of ascorbic acid in the

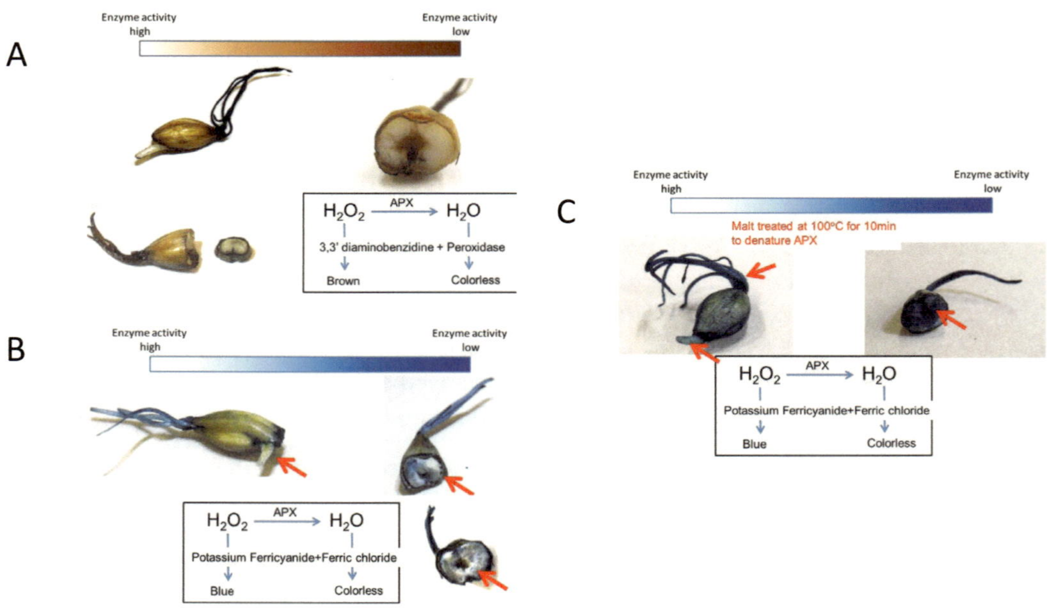

Figure 2. Detection of ascorbate peroxidase in sprouting barley (A) using 3,35′-diaminobenzidine as stain, (B) using potassium ferricyanide plus ferric chloride as stain, and (C) using potassium ferricyanide plus ferric chloride as stain on grain heated at 100°C for 10 min.

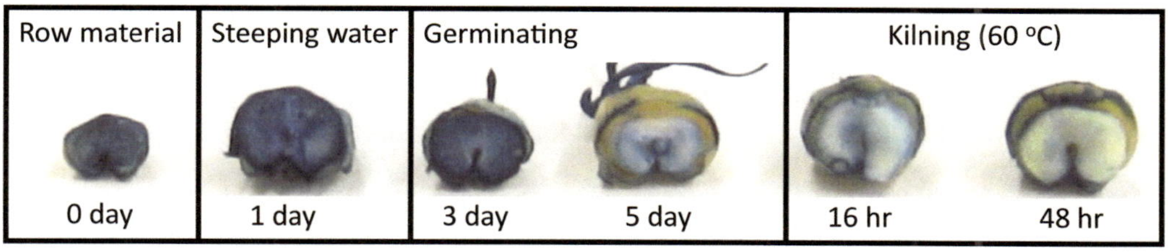

Figure 3. Detection of ascorbate peroxidase using potassium ferricyanide plus ferric chloride in half kernels of barley (A) sprouted for 1–6 days (B) in barley, during steeping, during germination, and after light kilning.

embryo, mid-grain, and distal regions of germinating barley is shown in **Figure 4A** and **B**. Each of the enzyme and ascorbic acid was detected in all three segments of the kernels after 2 days of germination. It is noteworthy that the mid-region has the highest levels of the enzyme and ascorbic acid. Kanauchi and Bamforth [19] inferred that these high levels reflect the development of enzyme in the acrospire growing under the husk, together with the development of the enzyme in the endosperm, presumably as a function of synthesis in the aleurone and rapid distribution into the cells of the starchy endosperm.

Synthesis of enzyme was attempted using inhibitors of RNA and protein and induction of a substrate. Inhibitors of RNA and protein synthesis, such as actinomycin D (80 μg/mL), 6-methylpurine (0.1 mM), puromycin (250 μg/mL), or cycloheximide (20 μg/mL), were added to barley during the development of ascorbate peroxidase. Actinomycin D and 6-methylpurine inhibited RNA synthesis. Furthermore, puromycin and cycloheximide

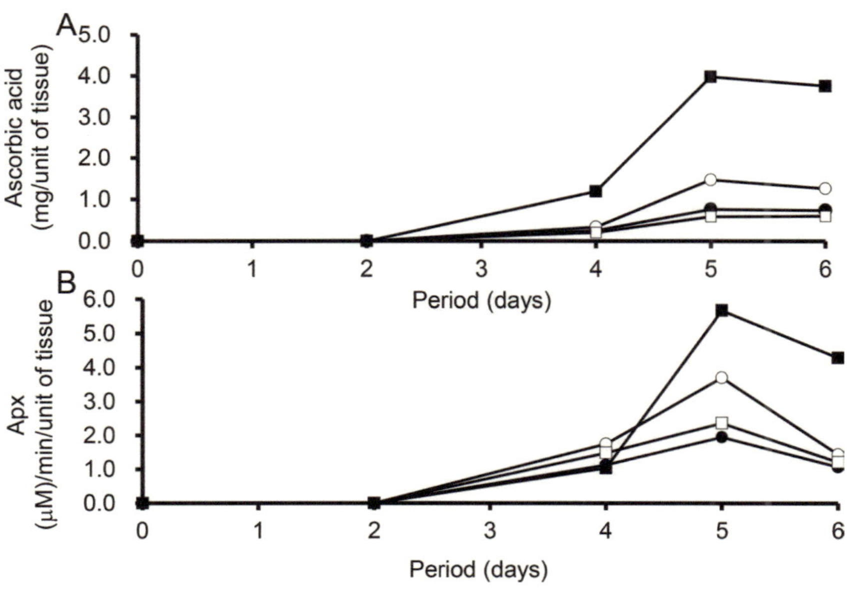

Figure 4. Incubation of barley slices (see **Figure 1**) in 10–5 M gibberellic acid solution: (A) levels of ascorbic acid; and (B) levels of ascorbate peroxidase. Embryo (•), endosperm 1 (○), endosperm 1 (□), and whole corn (■).

inhibited protein synthesis. Actinomycin D inhibited the development of new messenger RNA [24]. Puromycin did not inhibit enzyme development, but cycloheximide was a potent inhibitor. Kanauchi and Bamforth [19] concluded that the development of APX is a requirement for RNA and protein synthesis in the endosperm tissue.

The enzyme can also be induced by ascorbic acid and especially hydrogen peroxide and by a combination of the two. Results show that these are the active factors triggering the development of this enzyme (**Table 2**).

APX of crude extraction from malt was conducted by ammonium sulfate precipitate, chromatography using DEAE gel in column (25 mm × 300 mm), and size-exclusion column (10 mm × 350 mm, P-100 gel, Bio-Rad). It is apparently a monomer with relative molecular weight of approximately 26 kDa according to SDS-polyacrylamide gel electrophoresis. APX has an optimum pH of 5.5, with substantial activity at pH 5–7. Because the lower range pH encompasses the range of mash pH, the enzyme is reacted in mashing. However, it has very low capability to catalyze the removal of peroxide at pH 4–4.5, this condition meant that the enzyme reaction using an ascorbate for protection of beer against oxidation is extremely limited. Less than 50% of the activity survives heating at 50°C for 30 min. The enzyme is fundamentally destroyed in less than 10 min at 60°C (**Figure 5**).

Susceptibility of the enzyme to the chelating agents ethylenediaminetetraacetic acid (EDTA) and ethylene glycol tetraacetic acid (EGTA) suggests the necessity of a cation(s) for its full activity; but of the metal ions tested, only manganese exhibits any marked enhancement.

The enzyme is certainly inhibited by iron and copper, suggesting that the enzyme is inactivated by reactive oxygen species. Inhibition by azide suggests the presence of a functional heme group in the enzyme; indeed, heme has been shown to be the cofactor in this enzyme in other plants [25, 26]. Inhibition by iodoacetate indicates the functional presence of thiol groups, agreeing with others [25, 27]. Inhibition by N-bromosuccinimide (NBS) suggests a role for tryptophan and inhibition by benzenesulfonyl fluoride (BSF) would be consistent with a role for a serine group.

Two substrate kinetic analyses were conducted according to Dalziel (**Table 3**) [28], revealing that the Km values for ascorbic acid ($\phi 1/\phi 0$) and hydrogen peroxide ($\phi 2/\phi 0$) were 1.09 mM and 24.8 μM, respectively. Moreover, V_{max} ($1/\phi 0$) was 769 mM min^{-1}. The enzyme also had higher affinity for peroxide as opposed to ascorbate, which was true also for some other APX enzymes [26, 27].

	mU/10 slices	**Error (mU)**
Non	0.0	(+18.7)
Ultrafiltered extract	121.3	(+0.0)
5 mM H_2O_2	261.3	(+23.3)
5 mM ascorbic acid	46.7	(+39.7)
5 mM H_2O_2 + 5 mM ascorbic acid	280.0	(+14.0)

Table 2. Effects of hydrogen peroxide and ascorbic acid on the development of ascorbate peroxidase in barley slices.

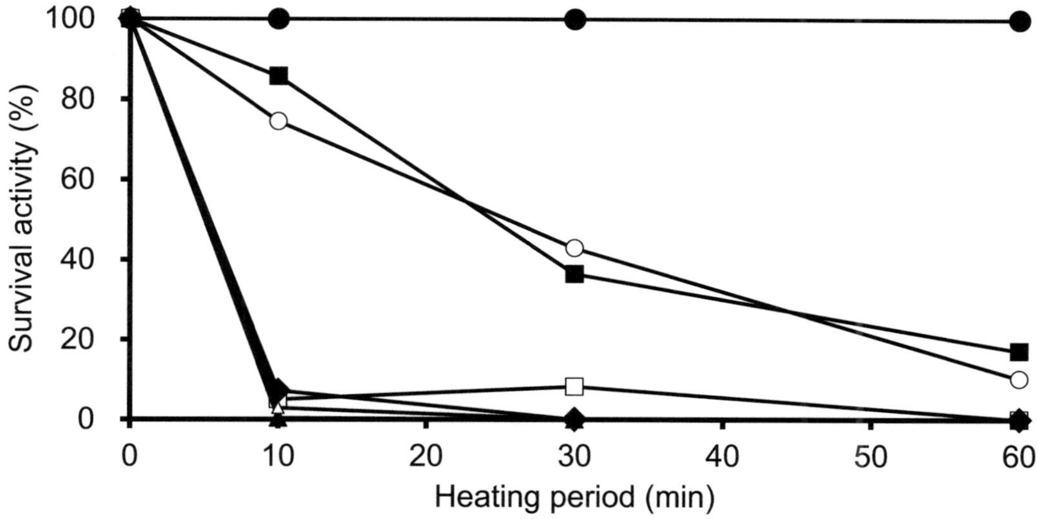

Figure 5. Heat tolerance of ascorbate peroxidase. Purified enzyme was heated at the temperatures indicated for 30 min before rapid cooling and subsequent assay. Symbols are 4°C (●), 40°C (○), 50°C (■), 60°C (□), 70°C (▲), 80°C (△), and 90°C (◆).

Ascorbate peroxidase was found in the acrospires and aleurones of germinating barley. Ascorbic acid and hydrogen peroxide induced the enzyme, which has a molecular weight of 23 kDa and a broad pH optimum, but which is sensitive to heat. Ascorbate peroxidase has very high affinity for hydrogen peroxide.

2.3. Ascorbic oxidase

Kanauchi et al. [29] reported the presence of ascorbic acid oxidase (AAO) and showed its properties.

Ascorbic acid oxidase (AAO EC 1.10.3.3) was found as "hexoxidase" in cabbage leaves in 1931 [30]. It catalyzes the following reaction.

$$2\ \text{L-ascorbate} + O_2 \rightarrow 2\ \text{Dehydroascorbate} + 2\ H_2O \qquad (3)$$

ϕ0 (S)	**ϕ1 (μM·S)**	**ϕ2 (μM·S)**	**ϕ12 (μM2·S)**	**ϕ1/ϕ0 (μM)**	**ϕ2/ϕ0 (μM)**	**ϕ12/ϕ2 (μM)**	**ϕ1ϕ2/ϕ12**
0.0013	1.46	0.03	57.93	1089.8	24.76	1744.7	0.000837

The φ parameters are calculated from the secondary plots that are developed as described in the Materials and methods section. φ0 is the intercept on the ordinate of the secondary plot of ordinate intercepts of the primary plot against the reciprocal of the second substrate concentration. φ2 is the slope of this line. φ1 is the ordinate intercept of the plot of primary plot slopes against the reciprocal of the second substrate concentration. φ12 is the slope of this line. 1/φ0 represents the true maximum velocity (V_{max}). φ1/φ0 equals the Km for the primary substrate (ascorbate). φ2/φ0 is the Km for the secondary substrate (hydrogen peroxide).

Table 3. Kinetic parameters of ascorbate peroxidase.

AAO has been widely observed in various plant and fungal tissues [31–36]. Tamas et al. [37] reported AAO in germinating barley seed with the agent causing substantial inhibition of rootlet growth. Honda [38] reported the enzyme's presence in barley roots and found it to be associated with cell walls. Kanauchi et al. [29] reported the existence in malt of a related enzyme: ascorbate peroxidase. Results show that this peroxidase has very high affinity for hydrogen peroxide and that the enzyme might have a valuable role in removing reactive oxygen species. However, the enzyme is heat-sensitive and therefore would not survive well during mashing.

Barley was germinated according to Hoy et al. [14]. AAO activity was assayed based on the measurement of the oxidation of ascorbate by the decrease in absorbance at 265 nm (e = 14 mM cm^{-1}) at 25°C [39]. One unit of enzyme catalyzes the oxidation of 1 mM ascorbic acid (AA) per minute.

AAO is not present in ungerminated barley, but starts to be synthesized immediately upon steeping (**Figure 6**). It reaches a maximum level of activity early in germination, thereafter decreasing to a low but finite level at the end of germination. AAO was purified as follows. Crude extract was applied to a column of CM support gels. The protein was eluted using a 0–1 M linear gradient of sodium chloride flowing at 1.5 mL min^{-1}. Fractions containing AAO were collected and precipitated using 80% saturation of ammonium sulfate. The precipitate was redissolved in citrate-phosphate (pH 7.0) and was then applied to the size-exclusion column. The eluent was citrate-phosphate.

After fractionation, crude extracts of malt by cation exchange chromatography revealed two peaks of AAO activity (data not shown). They are designated as AAO I and AAO II. Further chromatography of these peaks on Bio-Gel P100 (Bio-Rad) indicated that AAO I was of higher molecular size than AAO II. This characteristic was confirmed using polyacrylamide gel electrophoresis. Molecular weight estimates for the two enzymes are, respectively, 25–27 and 6–9 kDa. Two AAO enzymes were isolated from barley grains that are extremely different from any previously reported AAO [32–36]. Both enzymes are of much lower molecular size than previously

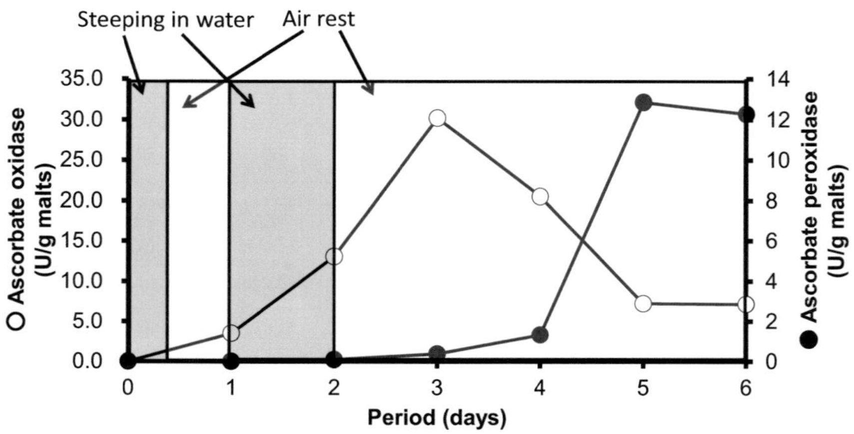

Figure 6. Levels of ascorbic acid oxidase and ascorbate peroxidase during steeping and germination of barley. Malt prepared at 15°C. Barley was sterilized using 1% calcium hypochlorite.

reported activities. AAO I had weakly cationic enzyme, with molecular weight of approximately 25 kDa. Furthermore, AAO II, which had a strongly cationic enzyme, had extremely low molecular weight of <10 kDa. The latter is one of the smallest enzymes ever reported: it is classifiable as a microenzyme [40, 41].

Optimum pH of AAO I and AAO II was pH 7.0. Furthermore, both activity pH had a broad pH range: AAO I has 66% relative activity; AAO II has 45% relative activity at pH 5.0. Furthermore, the optimum temperature of AAO I was 40°C, that of AAO II was 50°C. The enzyme displays slight activation by manganese and zinc. However, it is inhibited by copper, although AAO in most plants is generally described as an enzyme that is rich in copper. Iron, magnesium, and mercury inhibit them, especially AAO II. Furthermore, inhibition by chelating agents, such as EDTA and EGTA, is consistent with the need for a metal ion in the action of the enzymes. Inhibition by azide suggests the presence of a functional heme group in the enzyme. Inhibition by iodoacetate indicates the functional presence of thiol groups. Inhibition by N-bromosuccinimide suggests a role for tryptophan. Inhibition by benzenesulfonyl fluoride is consistent with a role for a serine group.

Kinetic parameters were calculated according to Dalziel [28] using the system shown in **Figure 7**. Substrate solutions (1 mL, 40°C) containing 0.0625–1.0 mM AA were introduced into

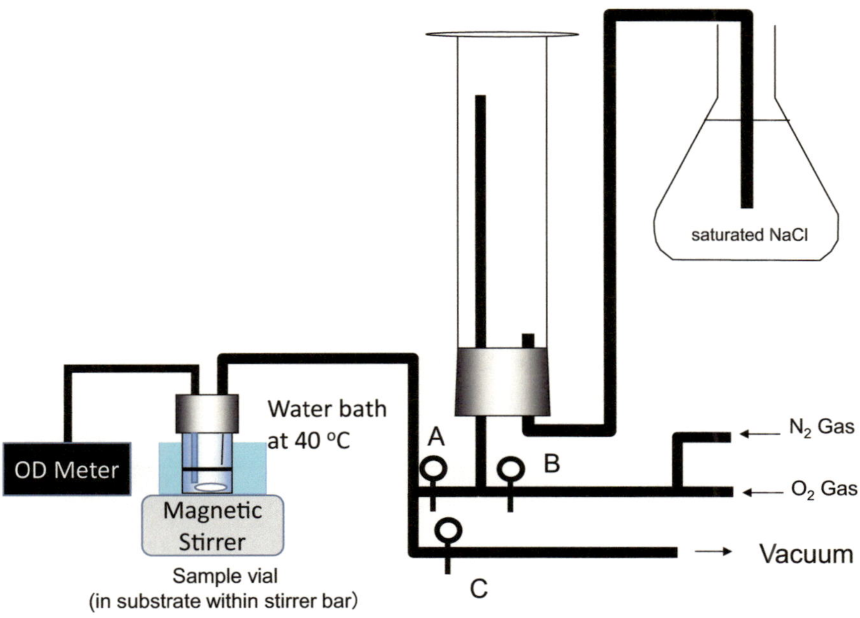

Figure 7. Apparatus for determining kinetic parameters for ascorbic acid oxidase. Air is removed from the vial using vacuum with stopcock C open. To adjust the oxygen concentration, a mixture of oxygen and nitrogen (e.g. 10 mL of O_2 and 90 mL of N_2) is transferred to the measuring cylinder filled with saturated NaCl solution via stopcock B. Upon closing B and opening stopcock A, the gas mixture is drawn vigorously to the vial from the cylinder. The operation is repeated three times to achieve a stable oxygen content, as measured using the dissolved oxygen meter. Enzyme is added to the substrate mixture by microsyringe. After reaction, 10 μL of the vial contents are transferred by microsyringe for measurement of absorbance at 265 nm.

a vial containing an electrode for measuring dissolved oxygen, which was then sealed with a rubber seal. A vacuum was applied in the vial. Then nitrogen gas, oxygen gas, and nitrogen-oxygen gas mixture (20–80% oxygen in nitrogen gas) were flushed successively through the vial. The operation was repeated three times. The enzyme solution was then added through the seal using a microsyringe. Decreased absorbance at 265 nm of the solutions was measured (Nano-Drop 2000; thermo Fisher Scientific Inc.).

Two substrate kinetic analyses (**Table 4**) revealed that AAO I can operate faster than [V_{max}] of AAO II. However, AAO II has much greater affinity for both substrates (lower Km values). The enzyme level declines as germination is prolonged. Actually, AAO II has much greater affinity (lower Km) for both substrates than AAO I does, although the latter displays a higher V_{max} value. The Km value for AA displayed by AAO II is comparable with that reported for AAO from other organisms [32–38]. Few other papers report a Km value for oxygen, but the value we have measured for AAO II is comparable to that reported for *Acremonium* spp. [36].

Impact of AA addition during mashing was conducted as follows. Milled pale malt (50 g) was mashed at 65°C in a water bath with 150 mg of AA and 150 mL of deionized water. The mashes designated for 0 min were filtered immediately upon mixing. Subsequent mashes were sampled periodically at 10, 20, 40, and 60 min. Mashes were filtered through cone filters into an ice bath. Wort samples were cooled to 4°C and analyzed as soon as possible. Density, pH, color, total polyphenols, and free thiols in wort samples were determined. Mashes were performed at 65°C in the presence of 5.7 mM ascorbic acid (AA).

This value is well more than the km value for AAO II. Furthermore, the enzyme would be expected to operate at a maximum rate at the start of mashing (**Table 5**). Adding AA had little impact on the specific gravity of the recovered wort. Although the pH of the mash was lowered by AA initially, it rose progressively during mashing. The pH of the control mash decreased.

The addition of AA led to markedly higher levels of polyphenol and thiols being measurable in the wort.

This result is consistent with reports of AA functioning: AAO consumed oxygen, which was used to oxidize polyphenols and thiols. There is generally also a lower color observed in

	Φ0 (s)	**Φ1 (mM s)**	**Φ2 (mM s)**	**Φ12 (mM^2 s)**	**1/Φ0 (s^{-1})**	**Φ1/Φ0 (mM)**	**Φ2/Φ0 (mM)**
AAO I	0.0012	0.0039	0.0129	0.0102	833	3.25	10.8
AAO II	0.0144	0.0051	0.0056	0.0105	69	0.35	0.39

The Φ parameters are calculated from secondary plots that are developed as described in the Materials and methods section. Φ0 is the intercept on the ordinate of the secondary plot of ordinate intercepts of the primary plot against the reciprocal of the second substrate concentration. Φ2 is the slope of this line. Φ1 is the ordinate intercept of the plot of primary plot slopes against the reciprocal of the second substrate concentration. Φ12 is the slope of this line. 1/Φ0 represents the true maximum velocity (V_{max}). Φ1/Φ0 equals the Km for the primary substrate. Φ2/Φ0 is the Km for the secondary substrate. Ascorbic acid is the primary substrate and oxygen is the secondary substrate.

Table 4. Kinetic parameters of ascorbic acid oxidase (AAO).

Time (min)	pH	Specific gravity	Polyphenol (mg/L)	Thiols (A430)	Color
Plus ascorbic acid					
0	5.145 ± 0.015	1.0267 ± 0.0038	177 ± 3	0.315 ± 0.032	1.97 ± 0.18
10	5.28 ± 0.01	1.082 ± 0.0014	294 ± 2	0.483 ± 0.009	4.9 ± 0.43
20	5.335 ± 0.005	1.0852 ± 0.0015	321 ± 3	0.476 ± 0.025	6.66 ± 2.63
40	5.385 ± 0.015	1.0978 ± 0.0005	347 ± 14	0.478 ± 0.004	7.47 ± 1.25
60	5.385 ± 0.015	1.1061 ± 0.0011	384 ± 1	0.471 ± 0.048	9.5
Control					
0	5.58 ± 0	1.055 ± 0.006	177 ± 2	0.071 ± 0.003	5.42 ± 0.04
10	5.575 ± 0.005	1.081 ± 0.012	189 ± 2	0.06 ± 0.023	8.31 ± 0.83
20	5.5 ± 0.006	1.087 ± 0.009	198 ± 1	0.07 ± 0.007	7.44 ± 0.51
40	5.49 ± 0.02	1.093 ± 0.003	212 ± 3	0.059 ± 0.012	9.45 ± 0.37
60	5.485 ± 0.005	1.103 ± 0.002	236 ± 2	0.054 ± 0.009	9.68 ± 0.81

Table 5. Effect of ascorbic acid addition during mashing.

small-scale laboratory mashes (with the exception of the 60 min measurement, which was perhaps a spuriously high value). This result is regarded as consistent with less polyphenol oxidation in the mashes in view of the extremely high affinity of AAO II for AA and oxygen, coupled with its thermotolerance.

There are some consequences to oxygen ingress in a mash, including possibilities of oxidation of unsaturated fatty acids, cross-linking of thiol-rich proteins, and oxidation of polyphenols with the production of color [12]. The addition of AA to mashes is expected to engender diminution in such effects. Bamforth et al. [4] anticipated increased measurable levels of such groups in mashes containing AA. An increased level of polyphenol surviving into wort and a decrease in the amount of color produced would be expected.

2.4. Oxalate oxidase

Oxalic acid can engender a range of problems in beer, including the blockage of dispensing pipes by beer stones, as well as turbidity and gushing [42, 43]. Importance of sufficient calcium to precipitate the material as calcium oxalate is notable. Few investigations have examined the origin of oxalic acid or of the enzymes that might produce or eliminate it during malting and brewing.

Oxalate oxidase (EC1.2.3.4) catalyses the conversion of oxalate into carbon dioxide and hydrogen peroxide. The hydrogen peroxide produced might serve some role in peroxidative cross-linking as part of cell wall restructuring [44]. The enzyme might also serve some role in countering the oxalate secreted by plant pathogens [45]. Oxalate oxidase is also known as germin [45, 46].

Kanauchi et al. [47] investigated oxalate oxidase of barley seed during malting and assessed its possible relevance in mashing. Barley was germinated according to Hoy [14]. The barley was germinated at 16°C for 6 days. The green malt was dried at 60°C for 16 h to approximately 5% moisture. The rootlets were not removed.

Oxalate oxidase was assayed according to the following principle [48].

$$\text{Oxalate} + O_2 \rightarrow 2CO_2 + H_2O_2(\text{catalyzed by oxalate oxidase}) \quad (4)$$

$$\text{MBTH} + H_2O_2 + \text{DMA} \rightarrow \text{Indamine dye (purple color)} + 2H_2O \; (\text{catalyzed by added peroxidases}) \quad (5)$$

where MBTH is 3-methyl-2-benzothiazolinone hydrazone; DMA is N,N-dimethylaniline.

Enzyme solution was mixed with oxalic acid solution (pH 4.0), MBTH solution, N,N-dimethylaniline, 3-methyl-2-benzothiazolinone hydrazone, ethylenediaminetetraacetic acid (EDTA) solution, and peroxidase enzyme solution. The standard curve was produced using hydrogen peroxide. One unit (1 U) of enzyme activity is defined as the amount of enzyme that catalyzes the oxidation of 1 μmol of substrate per minute under assaying conditions.

Its location in the grain remains as a matter for speculation. Oxalate oxidase was present in unmalted grain. Its activity increased slightly during germination (**Figure 8**). In germinated grain, 61% of the oxalic acid was in the rootlets, 32% in the acrospires, and the balance in the bran (husk/pericarp-testa/aleurone). Oxalic acid was not found in the starchy endosperm. It is noteworthy that more than half of the oxalate is located in the rootlets, which are removed

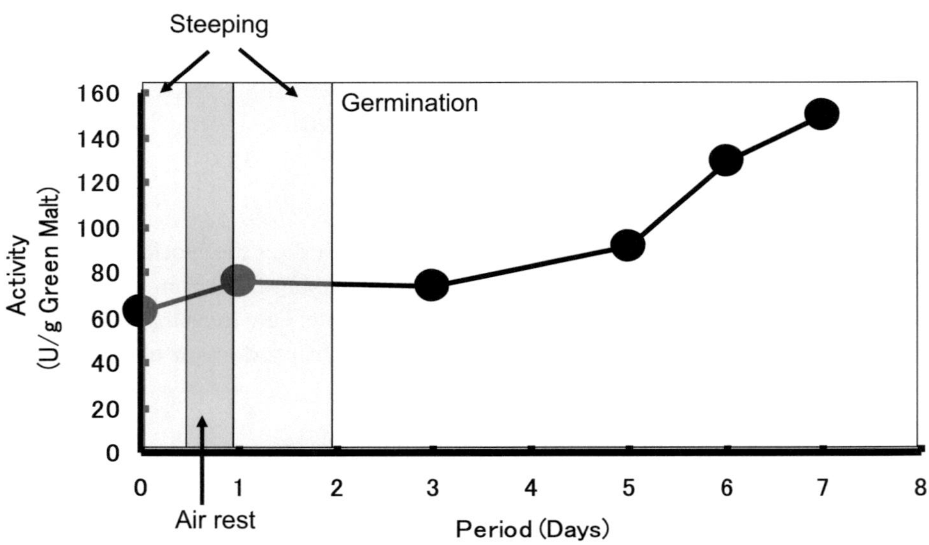

Figure 8. Oxalate oxidase activity of Starling barley during malting.

during malt production. Oxalic acid was not detectable in barley. Concentrations of oxalate in kernels increased during germination, although the amounts decreased near the end of germination. According to activity dyeing, the enzyme is present in the aleurone and the embryo, but not in the endosperm (**Figure 9**).

Ground-germinated grain was extracted in two volumes of citrate-phosphate buffer. After centrifugation of slurry, the extract was purified by precipitation using ammonium sulfate, a DEAE column, and a size-exclusion column. Consequently, the enzyme was partially purified using ammonium sulfate precipitation, ion exchange chromatography, and gel permeation chromatography, with a finishing specific activity of 44.6 U/mg protein (19.6-fold purified). The molecular weight of the enzyme was 58.5 kDa as determined from SDS-PAGE. Others reported an oxalate oxidase of rootlets of barley, which is a pentamer with subunit molecular weight of 25 kDa [49].

The enzyme has a pH optimum of approximately 4.0, but it displayed 90% of the maximum activity at pH 3 and 40% of the maximum activity at pH 8.0. Two substrate kinetic analyses, which showed km values for the oxalate was 0.1 mM, km values for oxygen was 0.46 mM. Vmax value for oxalic acid were 18.2 mM/min and its value for oxygen was 285.7 mM/min.

Several metal ions influenced the activity of oxalate oxidase (data not shown), such as manganese which slightly activated the enzyme. Generally, a manganese-claiming enzyme is a manganese-containing enzyme [50]. However, zinc and copper activated the enzyme to a greater degree. The enzyme was inhibited by cobalt, by iron, and to a lesser extent by magnesium. Inhibition by mercury, dithiothreitol, and iodoacetamide suggests that the enzyme needed free thiol groups for activity. This result was confounded by the observation that N-ethylmaleimide did not inhibit its activity. The enzyme was inhibited by azide. It was activated by flavin adenine dinucleotide (FAD), and reportedly might be a flavoprotein.

Kanauchi et al. [47] reported that barley kernels contain oxalate oxidase located in the living tissues of roots, but not in the starchy endosperm. Because the enzyme is active in a broad pH range, and because it has high heat tolerance, it was active during mashing, but it was less important than other oxidases for scavenging oxygen from mashes because of its low affinity for oxygen.

Figure 9. Location of oxalate oxidase in grain. Blue coloration denotes enzymic activity. The top photographs show germinated grain, whereas the bottom photograph shows unmalted grain.

3. Conclusion

Malt contains some enzymes or characteristics that degrade beer quality. This chapter presents descriptions of them. Four oxidations were investigated. The enzymes in malt have heat stability. Its optimum pH (pH 8.0) shows that the enzyme limits the reaction in the mash at pH 5.0. The activity of thiol oxidase lessens during malt storage. Results suggest that this is part of the explanation for why stored malt displays better wort separation than newly kilned malt.

Ascorbate peroxidase was identified in the acrospires and aleurones of germinating barley. Its synthesis was induced by the presence of ascorbic acid and hydrogen peroxide. It was found to have a molecular weight of 26 kDa and a broad pH optimum: pH 5–7. However, the enzyme lost 50% of its activity in 30 min at 40°C. The enzyme has very high affinity for hydrogen peroxide.

Ascorbic acid oxidase(AAO), which is an antioxidant agent as ascorbic acid, is developed in embryo tissues of barley during steeping and during the initial stages of germination. The molecular weights of the two AAO enzymes in malt are 27.4 and 6.4 kDa, respectively; their optimum pH is 7.0 and their optimum temperature is 40–50°C. Both enzymes are extremely heat tolerant and are capable of acting over a broad pH range. These two enzymes are expected to function during conversion temperatures of mashing.

Addition of ascorbic acid to mashes results in the survival of higher levels of polyphenol and thiols into wort and reduced color in that wort, commensurate with AAO preferentially consuming oxygen. Consequently, in adding ascorbic acid, oxygen is less available for other reactions, such as thiol oxidation and polyphenol oxidation in mashes.

Oxalate oxidase in barley kernels is located in living tissues of roots. The enzyme had active in a broad pH range, and it has high thermal tolerance. However, it is probably less important than other oxidases in scavenging oxygen from mash, because of its affinity for oxygen was low.

As explained herein, beer quality can be improved by the regulation of dissolved oxygen and oxidation enzymes, such as thiol oxidase, AAO of the endosperm cell walls in malt.

Author details

Makoto Kanauchi

Address all correspondence to: kanauchi@myu.ac.jp

Miyagi University, Sendai City, Japan

References

[1] Kanauchi M, Koizumi T, editors. Fermented Food Kodansha Scientific. Tokyo; 2012

[2] Kirin Company Limited. Available form: http://www.kirinholdings.co.jp/english/news/ [Accessed: 20 December 2016]

[3] Bamforth CW, Russell I, Stewart GG. Beer: A Quality Perspective. Burlington, MA: Academic Press; 2008

[4] Bamforth CW. Beer: Tap into the Art and Science of Brewing. New York: Oxford University Press; 2009

[5] Briggs DE. Malts and Malting. London, UK: Blackie Academic & Professional; 1998

[6] Muts GCJ, Pesman L. The influence of raw materials and brewing process on lautertun filtration. In: Proceedings of the EBC Monograph XI. Mafflier; 1986. pp. 25-35

[7] Moonen JHE, Graveland A, Muts GCJ. The molecular structure of gelprotein from barley, its behaviour in wort-filtration and analysis. Journal of the Institute Brewing. 1987;**93**:125-130

[8] Rennie H, Ball K. The influence of malt storage on wort separation. Journal of the Institute Brewing. 1979;**85**:247-249

[9] De Clerck J. A Textbook of Brewing. London: Chapman & Hall; 1957

[10] Baxter ED, Reeves SG, Bamforth CW. The effects of increased steeping temperatures on enzyme development in malt. Journal of the Institute Brewing. 1980;**86**:182-185

[11] Muller R. The formation of hydrogen peroxide during oxidation of thiol-containing proteins. Journal of the Institute Brewing. 1997;**103**:307-310

[12] Stephenson WH, Biawa J.-P, Miracle RE, Bamforth CW. Laboratory-scale studies of the impact of oxygen on mashing. Journal of the Institute Brewing. 2003;**109**:273-283

[13] Bamforth CW, Roza JR, Kanauchi M. Storage of malt, thiol oxidase, and brewhouse performance. Journal of the American Society of Brewing. 2009;**67**(2):89-94

[14] Hoy JL, Macauley BJ, Fincher GB. Cellulases of plant and microbial origin germinating barley. Journal of the Institute Brewing. 1981;**7**:77-80

[15] Ellman GL. Tissue sulfhydryl groups. Archives of Biochemistry and Biophysics. 1959;**82**: 70-77

[16] Clare DA, Bang WS, Cartwright G, Drake MA, Coronel P, Smunovic J. Comparison of sensory, microbiological, and biochemical parameters of microwave versus indirect UHT fluid skim milk during storage. Journal of Dairy Science. 2005;**88**:4172-4182

[17] Isaacs CE, Pascal T, Wright CE, Gaull GE. Sulfhydryl oxidase in human milk: Stability of milk enzymes in the gastrointestinal tract. Pediatrics Research. 1984;**18**:532-535

[18] Wu CK, Dailey TA, Dailey HA, Wang B.-C, Rose JP. The crystal structure of augmenter of liver regeneration: A mammalian FAD-dependent sulfhydryl oxidase. Protein Science. 2003;**12**:1109-1118

[19] Kanauchi M, Bamforth CW. Ascorbate peroxidase in malted barley. Journal of the American Society of Brewing Chemists. 2013;**71**(2):97-102

[20] Shigeoka S, Ishikawa T, Tamoi M, Miyagawa Y, Takeda T, Yabuta Y, Yoshimura K. Regulation and function of ascorbate peroxidase isoenzymes. Journal of Experimental Botany. 2002;**53**:1305-1319

[21] Bamforth CW, Martin HL. The development of β-glucan solubilase during barley germination. Journal of the Institute Brewing. 1981;**87**:81-84

[22] Manchenko GP. Handbook of Detection of Enzymes on Electrophoretic Gels. Boca Raton, FL: CRC Press; 1994

[23] Nakano Y, Asada K. Hydrogen peroxide is scavenged by ascorbate-specific peroxidase in spinach chloroplasts. Plant Cell Physiology. 1981;**22**:867-880

[24] Chrispeels MJ, Varner JE. Gibberellic acid-enhanced synthesis and release of α-ribonuclease by isolated aleurone layers. Plant Physiology. 1967;**42**:398-406

[25] Jones DK, Dalton DA, Rosell FI, Raven EL. Class I heme peroxidases: Characterization of soybean ascorbate peroxidase. Archives of Biochemistry and Biophysics. 1998;**360**:173-178

[26] Koshiba T. Cytosolic ascorbate peroxidase in seedlings and leaves of maize (Zea mays). Plant Cell Physiology. 1993;**34**:713-721

[27] Sharma P, Dubey RS. Ascorbate peroxidase from rice seedlings: Properties of enzyme isoforms, effects of stresses and protective roles of osmolytes. Plant Science. 2004;**167**:541-550

[28] Dalziel K. Initial state velocities in the evaluation of enzyme-coenzyme-substrate reaction mechanisms. Acta Chemica Scandinavica. 1957;**11**:1706-1723

[29] Kanauchi M, Simon KJ, Bamforth CW. Ascorbic acid oxidase in barley and malt and its possible role during mashing. Journal of the American Society of Brewing Chemists. 2014;**72**(1):30-35

[30] Szent-Györgyi A. On the function of hexuronic acid in the respiration of the cabbage leaf. Journal of Biological Chemistry. 1931;**90**:385-393

[31] Simpson BK, Nollet LML, Tordera F, Soottawat Benjakul S, Paliyath G, Hui YH. Food Biochemistry and Food Processing. John Wiley & Sons; 2012. p. 185

[32] Every D. Purification and characterization of ascorbate oxidase from flour and immature wheat kernels. Journal of Cereal Science. 1999;**30**:245-254

[33] De Tullio MC, Ciraci S, Liso R, Arrigoni O. Ascorbic acid oxidase is dynamically regulated by light and oxygen. A tool for oxygen management in plants? Journal of Plant Physiology. 2007;**164**:39-46

[34] Bezerra Carvalho L, Lima CJ, Medeiros PH. Ascorbate oxidase from Cucurbita maxima. Phytochemistry. 1981;**20**:2423-2424

[35] Porto TS, Porto CS, Cavalcanti MT, Filho JL, Perego P, Porto AL, Converti A, Pessoa A. Kinetic and thermodynamic investigation of ascorbate oxidase activity and stability of a Cucurbita maxima extract. Biotechnology Progress. 2006;**22**:1637-1642

[36] Itoh H, Hirota A, Hirayama K, Shin T, Murao S. Properties of ascorbate oxidase produced by Acremonium sp. HI-25. Bioscience, Biotechnology, and Biochemistry. 1995;**59**: 1052-1056

[37] Tamas L, Bocova B, Huttova J, Mistrik I, Olle M. Cadmium-induced inhibition of apoplastic ascorbate oxidase in barley roots. Plant Growth Regulation. 2006;**48**:41-49

[38] Honda SJ. Ascorbic acid in barley roots. Plant Physiology. 1955;**30**:174-181

[39] Oberbacher MF, Vines HM. Spectrophotometric assay of ascorbic acid oxidase. Nature. 1963;**197**:1203-1204

[40] Mattey M, Simoes D, Brown A, Fan X. Enzymes with a low molecular weight. Acta Chimica Slovenica. 1998;**45**:45-57

[41] Monti D, Riva S. Natural and artificial microenzymes: Is it possible to have small and efficient biocatalysts? Biocatalysis and Biotransformation. 2001;**19**:251-266

[42] Burger M, Becker K. Oxalate studies on beer. Proceedings of the American Society of Brewing Chemists. 1949;**7**:102-115

[43] Burger M, Glenister PP, Becker K. Oxalate studies in beer. II. Supplementary work and observations on calcium-oxalate relationships. Proceedings of the American Society of Brewing Chemists. 1956;**14**:169-178

[44] Davoine C, Le Deunff E, Ledger N, Avice J.-C, Billard J.-P, Dumas B, Huault C. Specific and constitutive expression of oxalate oxidase during the ageing of leaf sheaths of ryegrass stubble. Plant Cell Environment. 2001;**24**:1033-1043

[45] Dunwell JM, Khuri S, Gane PJ. Microbial relatives of the seed storage proteins of higher plants: Conservation of structure, and diversification of function during evolution of the cupin superfamily. Microbiology and Molecular Biology Reviews. 2000;**64**:153-179

[46] Lane BG, Dunwell JM, Ray JA, Schmitt MR, Cuming AC. Germin, a protein marker of early plant development, is an oxalate oxidase. Journal of Biological Chemistry. 1993;**268**:12239-12242

[47] Kanauchi M, Milet J, Bamforth CW. Oxalate and oxalate oxidase in malt. Journal of the Institute Brewing. 2009;**115**(3):232-237

[48] Laker MF, Hoffman AF, Meeuse JD. Spectrophotometric determination of urinary oxalate with oxalate oxidase prepared from moss. Clinical Chemistry. 1980;**26**:827-830

[49] Kotsira VP, Clonis YD. Oxalate oxidase from barley roots: Purification to homogeneity and study of some molecular, catalytic, and binding properties. Archives of Biochemistry and Biophysics. 1997;**340**:239-249

[50] Woo E-J, Dunwell JM, Goodenough PW, Marvier AC, Pickersgill RW. Germin is a manganese-containing homohexamer with oxalate oxidase and superoxide dismutase activities. Nature Structure and Molecular Biology. 2000;**7**:1036-1040

Chapter 3

Barley (*Hordeum vulgare* L.) Improvement Past, Present and Future

Nermin Gozukirmizi and Elif Karlik

Additional information is available at the end of the chapter

http://dx.doi.org/10.5772/intechopen.68359

Abstract

Barley has been cultivated for more than 10,000 years. Barley improvement studies always have the privilege of the breeders and scientists. This review is expected to provide a resource for researchers interested in barley improvement research in terms of mutation breeding, tissue culture, gene transfers, gene editing, molecular markers, transposons, epigenetic, genomic studies and system biology. We aimed to discuss some important and/or recent studies and improvements about barley for understanding the factors responsible for converting barley plants into the superior cereals, which occurred through gene transfers, gene editing and molecular breeding, which is important and could help us enhance the current pool of cultivated barley species to provide enough material for the future.

Keywords: barley improvement, *Hordeum vulgare* L., genetic research, genomics research, complex trait

1. Introduction

Cultivated barley (*Hordeum vulgare* L.) is the fourth important annual cereal crop from the family of Poaceae after wheat, rice, maize and is consumed as feed for livestock and, food—either pure or combined with other cereals in the form of porridge, sattu (roasted barley), breakfast foods and chapattis [1] and, most importantly, is also used for brewing malts. Barley, which is also an excellent model plant for biochemists, physiologists, geneticists and molecular biologists, is one of the world's earliest domesticated and most important crop plants [2]. According to world statistics, its production in 2015 was 148.78 million tons, where Turkey's contribution was 4,750,000 metric tons [3]. Barley is a self-pollinating diploid with $2n = 2x = 14$ chromosomes. Moreover, it has two-rowed and six-rowed types, according to spike morphology [4]. The barley genome project is completed by the International Barley

Genome Sequencing Consortium [5]. It has 26,159 genes and large haploid genome of 5.1 gigabases (Gb), approximately 84% of the genome is comprised of mobile elements or other repeated structures. Ease of growth under laboratory conditions, and tissue cultures facilitate the development of gene transfer and gene editing technologies, although research on barley genome and system biology is progressing.

Barley has been cultivated for more than 10,000 years [6]. In former times, the Sumerian and Babylonian cultures utilized barley grains as currency. Barley improvement studies always have the privilege of the breeders and scientists. Barley is a short season, early maturing grain with a high-yield potential, and may be found on the fringes of agriculture in widely varying environments, often on the fringes of deserts and steppes or at high elevations in the tropics, receiving modest or no inputs [7]. Wide genetic variation of barley has generated cultivars that are tolerant to stress environments such as cold, salinity, drought and alkaline soil [8]. It is possible to cultivate barley in extensive ecological range. This adaptive genetic diversity against abiotic and biotic stresses indicates the potential of barley to develop stress resistant cultivars. The main objective of barley breeding programmes is enriching yield and grain quality. Improvement studies are also based on producing varieties resistant to biotic (pathogens, fungal, viral and other organisms) and abiotic stresses (e.g. drought, salt, cold and heat) [9]. Identifying and understanding the genetics basis of stress tolerance mechanisms in crops is fundamental to develop new varieties with more stress tolerant characters [10].

Barley is an economically important crop plant, the fourth cereal worldwide in terms of the planting area, utilized almost 60% as animal feed, around 30% for malt production, 7% for seed production and only 3% for human food [11, 12]. In recent years, the malt derived from the germinated barley is the key material for the malting which represents the most economically favourable application for beer brewing [13]. However, to enhance the germination and malting quality of barley, addition of malting additivives during the malting is strictly controlled cause of food safety and environmenal pollution. Improvement of barley cultivars for the malting may be the most economical approach to improve malt quality. As a result, identifying and understanding the genetics basis of barley is fundamental to develop new varieties with more properties [14]. Also nowadays, barley has numerous advantages in food industry due to its high content of bioactive compounds such as β-D-glucan, tocopherols, tocotrienols and phenolics such as benzoic and cinnamic acid derivatives, proanthocyanidins, quinones, flavonols, chalcones and flavones [15, 16]. The studies showed that β-D-glucan is regarded as a significant function of preventing various diseases such as diabetes, cardiovascular diseases, hypertension and others [17].

Barley is one of the most genetically diverse cereals which is categorized as spring or winter types, two-rowed six row, hulled or hulless by the presence or absence of hull tightly adhering to the grain, and malting or feed by end-use type. Therefore, breeding programmes depend on high level of genetic diversity which provides a significant opportunity for achieving progress. Specific traits may be introgressed in back-crossing studies by hybridisations between high-yielding cultivars and wild barley in conventional breeding programmes [18]. However, mutation breeding is also important for widening variation to develop new cultivars. Herman Nilsson-Ehle and Ake Gustafsson, and even L. J. Stadler have performed induced mutation studies on barley, and then Stadler have published his data in 1928. In 1953, the 'Group for theoretical and applied mutation research' was established by the Swedish Government. The aim of their study was the investigation of basic research problems in order to effect and improve methods for breeding programmes [19]. Both radiation and chemical mutagenesis have been separately used to increase the numbers of barley cultivars which may have desirable traits. 'Golden promise', which is the most popular malting barley, was produced by radiation mutagenesis [20]. In Turkey, mutation breeding programme has been started by Bilge et al. with collaboration of Agricultural Research Institutes [21, 22]. They treated barley seeds with radiations (X and gamma rays) and chemical (ethyl alcohol, streptomycin, terramycin, penicillin G, sodium cyanide and ethyl methane sulfonate solutions) mutagens and observed different traits such as chlorophyll deficiency, large-eared, high-yielding, thick-stemmed, dwarf and early-heading in M1. Today, use of mutation breeding generally continuing at targeted level will be discussed by new technologies.

In this review, we summarize the history of barley improvement research in terms of mutation breeding, tissue culture, gene transfers, gene editing, molecular markers, transposons, epigenetic, genomic studies and system biology. We aimed to discuss some important and/or recent studies and improvements about barley for understanding the factors responsible for converting barley plants into the superior cereals, which occurred through gene transfers, gene editing and molecular breeding, which is important and could help us enhance the current pool of cultivated barley species to provide enough material for the future.

2. Barley molecular markers

Plant breeders have been used with phenotypic traits for selection of desirable traits due to habits, disease resistance, yield or quality to develop new cultivars. Two major strategies have been utilized to select desirable traits which are classical breeding and molecular breeding. The development and use of molecular markers for the detection and exploitation of polymorphism have been playing a significant role in plant breeding studies. Molecular plant breeding utilizes two major approaches, marker-assisted selection (MAS) and genetic transformation, to produce new varieties with desirable characteristics [23, 24]. MAS is a process that uses molecular markers to increase crop yield, quality and tolerance to biotic or abiotic stresses [25]. The choice of marker systems is a significant part of plant breeding cause of the requirements according to the conditions and resources. In the last two decades, molecular markers such as restricted fragment length polymorphism (RFLP), randomly amplified polymorphic DNA (RAPD), amplified fragment length polymorphisms (AFLPs), simple sequence repeats(SSR), inter-simple sequence repeats (ISSR), expressed sequence tags (ESTs) and single-nucleotide polymorphisms (SNPs), transposon-based markers (IRAP, iPBS) have been used as genetic markers for measuring the genetic differences existing in the genomes [26–30]. Development of next-generation sequencing technologies opened new opportunities for the development of sequence-based markers. Today, we have new markers which are not fragment-based but are sequence-based. Medium and high density arrays are available for barley. The choice of utilized marker methods has shifted from the first and second generation markers such as RFLPs, RAPDs, microsatellite and AFLPs to third and fourth generation markers including DArTs, TAMs, RADs and CNVs/PAVs which are demonstrated in **Table 1**

Marker types used	Aim	Results	Reference
RFLP	Construction of an RFLP map of barley	Genetic and physical mapping achieved	[37, 38]
RAPD	Analyses genetic variations in barley	Cultivar certification achieved	[56]
	Cultivar discrimination	Cultivar and hybrid certification achieved	[57]
AFLP	AFLP markers linked to water stress tolerant and sensitive bulbs of barley	AFLP markers was identified in two barley genotypes (tolerant and sensitive)	[41]
SSR	Construction of a SSR consensus map of barley	149 mapped SSR markers on Australian varieties are presented in the form of a consensus map, SSRs proved to be adaptable to several technologies	[44, 45]
IRAP and iPBS	Callus age and retrotransposon	Tissue culture conditions and callus age affected *Sukkula* retrotransposon movements, and all individuals did not present the same effect	[51]
SNP	Utilization of the BOPA1 assay to explore SNPs in geographically matched landraces and wild accessions collected	Of the 1536 SNPs represented on BOPA1, 1301 mapped SNPs	[58]
CAPS	Comparison of SNP and CAPS markers application in genetic research in wheat and barley	Results supporting the development of different strategies for the application of effective SNP and CAPS markers in wheat and barley	[59]
CNV	The prevalence of copy number variation (CNV) and its role in phenotypic variation in domesticated barley cultivars and wild barleys	Levels of CNV in the wild accessions were found to be higher than cultivated barley. CNVs are enriched near the ends of all chromosomes except 4H. CNV affects 9.5% of the coding sequences represented on the array and the genes affected by CNV	[55]
cDNA-AFLP	Development of molecular markers linked to barley heterosis	Five transcript-derived fragments (TDFs) showed significant effects on heterosis	[60]
DArT	94 Czech malting barley cultivars identification	DArT-based dendrogram was established	[61]
	Genome-wide association studies of agronomic and quality traits in a set of German winter barley (*Hordeum vulgare* L.) cultivars using DArT	A set of about 100 winter barley (*Hordeum vulgare* L.) cultivars, comprising diverse and economically important German barley elite germplasm was analysed	[62]
	QTL loci effecting kernel length	LEN-3H and LEN-4H could be used for improve kernel length	[63]
SLAF-seq and whole-genome shotgun	semi-dwarf gene *ari-e* from Golden Promise	Specific-length amplified fragment sequencing (SLAF-seq) with bulked Segregant analysis (BSA) to develop SNP markers, and (2) the whole-genome shotgun sequence to develop InDels. Both SNP and InDel markers were developed in the target region	[64]
Restriction site associated DNA (RAD) sequencing	SNP-based high density genetic map and mapping of *btwd1* dwarfing gene in barley	The SNP-based high-density genetic map developed and the dwarfing gene *btwd1* mapped	[65]

Marker types used	Aim	Results	Reference
InDel markers	Development of InDel markers	High-density InDel markers with specific genome locations were developed with 6976 molecular markers (SSRs, DArTs, SNPs and InDels) integrated into single barley genetic map	[66]

Table 1. Molecular markers used in barley research.

[31–34]. Next-generation breeding technologies are now effectively used for the establishment of genotypic and phenotypic relations [35]. Future barley varieties are designed with crop model ensemble [36].

Restriction fragment length polymorphism (RFLP) marker system has been used as a measure of genetic diversity for mapping studies in barley [37, 38]. Genetic relationships among 21 barley accessions (17 of *H. bulbosum* L. and 4 of *H. vulgare* L.) have been investigated by Okumus and Uzun [39] have successfully produced 111 RAPD markers. Combination of bulked segregant analysis and RAPD primers has been used to identify molecular markers linked to crown rust resistance gene *Rpc1* in barley [40]. Another molecular marker technique AFLP has been utilized for linkage studies and evolution of barley [41–43]. 149 simple sequence repeats (SSRs) or microsatellite markers have been constructed in the form of a consensus map by using 12 barley populations [44, 45]. SSR markers have been utilized for the selection of *Rym4*/*Rym5* locus conferring resistance to the barley mosaic virus complex in barley. The polymorphic SSR marker QLB1 was found to be co-segregated with *Rym4*/*Rym5* locus which also used to develop for the high-resolution map [46]. Other marker methods used in plant breeding are transposable elements-based marker systems such as inter-retrotransposon amplified polymorphism (IRAP), retrotransposon-microsatellite amplified polymorphism (REMAP) and inter-primer binding site amplification (iPBS) to identify retrotransposon markers linked to traits. Our group has been using IRAP and iPBS marker techniques to determine retrotransposon insertion patterns, movements of transposons, somaclonal variations, and callus aging. Our results showed that callus culture conditions have activated *BARE-1* and *Nikita* elements [47–50]. Movements of the non-autonomous retrotransposon *Sukkula* were investigated by Kartal-Alacam et al. [51] in barley. Recently, IRAP technique is also utilized to assess the genotoxicity of some drugs such as epirubicin [52] and amiprophos-methyl [53].

Genome- and chromosomal-level genetic structures are really important for the investigation of the evolution, adaptation and spread of the crops. Therefore, single-nucleotide polymorphism (SNP) platforms, which are used to assess the evolution of barley, are a key tool in the development of farming. Russell et al. [54] utilized the barley oligonucleotide pool assay 1 platform (BOPA1, composed of 1536 SNPs) to compare 448 accessions genome-level genetic structures, 317 of landrace material and 131 of wild barley, and observed that significant chromosome-level differences diversity between landrace and wild barley types was around genes known to be involved in the evolution of cultivars. Fourteen barley genotypes (eight cultivars and six wild barleys) have been utilized to explore copy number variations (CNV) by using comparative genomic hybridization. The study showed that CNVs were enriched near the ends of all chromosomes except 4H and affected 9.5% of the coding sequences represented on the array [55].

3. Barley tissue cultures and gene transfers systems

Plant tissue culture, which provides convenience for plant propagation and manipulation, is based on growing plant cells, tissues or organs isolated from the mother plant, on artificial media [67]. It is required to regenerate *in vitro* whole transgenic plants by using cells, tissues or a single cell cultured on a nutrient medium in a sterile environment [68]. Regeneration ability in barley depends on the donor plant material, genotype, media and environment [69–71]. One significant limitation of barley transformation is still the poor regeneration potential of modern cultivars. However, several studies have been conducted to improve tissue culture techniques to increase regeneration rates [72]. From past to today, various tissue culture protocols have been developed by using immature embryos [73–80], mature embryos [81–87], apical meristems [88–90], anthers [91–94], microspores [95–97], ovaries [98, 99], cell suspensions [100–104], protoplasts [105], coleoptile tissue [106] and leaf base segments [90, 107].

The improvement of barley through genetic transformation and *in vitro* methods requires the development of reliable, efficient and reproducible plant regeneration systems (**Table 2**) [70, 108, 109]. The plant regeneration capacity is affected by the genotype of donor plants, growth characteristics of induced calluses, the composition of the media, including growth regulators [110, 111]. Tissue cultures of barley are mainly based on the optimization of callus induction [112], regeneration [71, 113] and transformation [99], understanding of tissue culture response [114], detection, evaluation and elimination of somaclonal variation [81, 94, 115–118]. The use of mature embryos has a great advantage compared to other systems such as protoplast and cell suspensions. For barley tissue culture, mature embryos represent ideal system because of higher germination and regeneration rates by somatic embryogenesis from cultured mature embryos of barley [87]. Phytohormones are also crucial to setting optimal tissue culture conditions to produce undifferentiated callus tissue from differentiated tissues such as an embryo [119].

Callus formation, which is a dedifferentiation of single cells or tissue explants, offers the great opportunity for investigation of *in vitro* selection production of genetic variations [120–124]. The regeneration of plants from callus of barley has a great potential to produce new lines in breeding improved barley cultivars [125, 126]. The type of auxin, 2,4-dichlorophenoxyacetic acid (2,4-D), is the most used growth regulator for callus induction [123, 127, 128]. 2,4-D have been utilized to induce embryogenic callus together with or without cytokinins such as zeatin or 6-benzylaminopurine (6-BAP). Moreover, the influences of 2,4-D, Dicamba (3,6-dichloro-O-anisic acid), Picloram (4-amino-3,5,6-trichloropicolinic acid) or 2,4,5-T (2,4,5-trichlorophenoxyacetic acid) have been investigated on the induction of embryogenic callus. It was found that Dicamba significantly increased the regeneration through somatic embryogenesis [78, 111, 129–131]. However, callus quality depends

Culture type	Aim	Results	Reference
Immature embryos	Tissue culture and plant regeneration from immature embryo explants	Regeneration of plantlets was obtained for 19 of the 20 genotypes approximately 4 months after culture initiation	[75]
	Evaluation 9 barley cultivar for *in vitro* culture response	For each character, there were significant differences between genotypes, between 2,4-D concentrations and also significant genotype × medium interactions	[136]
	Evaluation of 10 Canadian barley genotypes for in vitro culture response	Fertile plants were regenerated	[137]
	Callus induction and regeneration at Czech cultivars	The callus formation frequency and number of green regenerants were influenced significantly both by genotype and auxin	[123]
	Callus induction and regeneration at Nordic cultivars	Regeneration of many plants from the same callus over long periods of time and makes available highly efficient regeneration protocols	[138]
Mature embryos	Tissue culture establishment and plantlet regeneration	Plantlets regenerated both via organogenesis and somatic embryogenesis	[139]
	Tissue culture and plant regeneration at Indian cultivars	Multiple shoot induction and plantlet regeneration in Indian cultivar of barley	[140]
Anther culture	Possible effect of copper during anther culture in barley	The positive influence of copper sulphate was characterized by an increase of microspore survival during anther culture	[141]

Table 2. Tissue culture and plant regeneration studies in barley.

on barley genotypes [125, 132]. And also, it has been reported that the most barley cultivars produced friable and translucent callus [122, 125].

Somatic embryogenesis, which is defined as a process by which haploid or diploid somatic cells develop into structure that resembles zygotic embryo, is an important tool for large scale vegetative propagation. Somatic embryos are bipolar structures without any vascular connection with the parental tissue and these structures can differentiate either directly from the explants without an intervening callus phase or indirectly after a callus phase. Immature embryos have a great potential to produce somatic embryos through embryogenic callus [133]. Marthe et al. [134] have investigated transformation efficiency for more than 20 barley cultivars by using immature barley embryos, and they found that the transformation efficiency of cv 'Golden Promise' was still higher than any other cultivar tested. Another study conducted by Hisano et al. [135] showed that callus derived from immature embryos of 'Golden Promise' had the highest ratio of regeneration of green shoots comparing with 'Haruna Nijo' and 'Morex'.

Since 1990s, genetic engineering of plants is a powerful research tool for gene discovery and function to investigate genetically that controlled traits have provided great opportunities to introduce agronomically useful traits. The first report on stable barley transformation via direct DNA-transfer methods has been established by Lazzeri et al. [105]. Tingay et al. [142] were first reported *Agrobacterium*-mediated gene transfer protocol to barley using immature embryos (IEs). Since then, numerous protocols for barley transformation have been developed with the contribution of technical improvements based on immature embryos or androgenetic pollen cultures or isolated ovules as gene transfer targets [99, 143–146]. Gurel and Gozukirmizi [147] optimized the transformation parameters for efficient and successful genetic transformation of mature barley embryos. They defined the optimal combination of electroporation and electroporated mature embryos with β-glucuronidase (*Gus*) and neomycin phosphotransferase II (*nptII*) genes. The frequency of transformants was generally not very high between 1.7 and 7.0% of the immature embryos infected via *Agrobacterium*. On the other hand, it has been recently reported that the frequency is around 25% or higher [148–150]. Although the transformation frequency is lower, immature embryos still remain the target tissue of choice in barley [148, 151].

The most of barley transformation studies have been performed to confer biotic (fungal and viral resistance) and abiotic (herbicide, drought and salinity, etc.) resistance, to facilitate brewing and digestibility, to alter protein composition and for molecular pharming [152]. Some of those methods have been established in **Table 3**. Yeo et al. [153] developed

Gene transfer type	Aim	Results	References
Biolistic transformation system	Target tissues such as immature embryos, embryos derived callus and microspore derived callus	Successful transformation	[156]
	Immature embryos and microspore-derived cultures	Successful transformation	[157]
	Transformation of recalcitrant species	Successful transformation	[158]
	Pre cultured immature embryos	Molecular analysis of T1 generation plantlets revealed the amplification of selectable marker hptII gene in the progeny	[159]
Agrobacterium-mediated transformation	Immature embryos	Successful transformation	[142]
	Shoot apices	Successful transformation	[160]
	Optimization of gene transfer immature embryos	Transformation efficiencies 2.6–6.7%	[145]
	Young ovules	Successful transformation	[161]
	Microspores	Successful transformation	[147]
	Optimization of gene transfer immature embryos	25 % transformation efficiency	[148]
	Mature scutellum	Successful transformation	[162]
	Immature embryo-derived callus cultures	Improve T-DNA transfer in monocotyledon transformation procedures	[163]
	Mature embryos	Successful transformation	[164]
Tissue electroporation	DNA transfer into mature embryos of barley via electroporation	Successful transformation	[147]

Table 3. Gene transfer research on barley.

'Golden SusPtrit' which is a barley line combining SusPtrit's high susceptibility to non-adapted rust fungi with the high amenability of Golden Promise. They generated a double haploid (DH) mapping population (n=122) by crossing SusPtrit with Golden Promise to develop the 'Golden SusPtrit'. SG062N was found the most efficiently transformed DH line with 11–17 transformants per 100 immature embryos. To protect barley from the effects of stress-produced reactive carbonyls, which is accumulated by reactive oxygen species in the plant cells, an *Agrobacterium*-mediated transformation was carried out using the *Medicago sativa* al dose reductase (MsALR) gene by Nagy et al. [154]. Their results demonstrated that this technique could be applied for the detection of cellular stress, and also found that targeting of MsALR into the chloroplast has also resulted in increased stress tolerance. In addition to these studies, Han et al. [155] reported that a construct containing full-length of HvGlb2 cDNA encoding barley (1,3;1,4)-β-glucanase isoenzymes EII under the control of a promoter of barley D-Hordein gene Hor3-1 was introduced into barley cultivar Golden Promise via *Agrobacterium*-mediated transformation. High content of (1,3;1,4)-β-D-glucan of barley grains is considered as an undesirable factor effecting malting potential, brewing yield and feed utilization. They showed that over-expression of (1,3;1,4)-β-glucanase led to an increase in the thousand grain weight. Also, manipulating expression of (1,3;1,4)-β-glucanase EII could control the β-glucan content in grain with no apparent harmful effects on grain quality.

4. Genomic studies on barley

The genetic revolution of the past decade has greatly improved our understanding of the relationships between genetic and phenotypic diversity with a resolution that has never been reached before. The development of next generation sequencing (NGS) technologies has increased accuracy and decreased costs. Sequencing or re-sequencing of reference genomes and also new varieties allow the identification of numerous numbers of markers, allelic diversities and have changed our insight of genome organization and evolution. The sequencing of crop genomes provided evidences for plant origin and evolution; genome duplications, re-arrangements; adaptations and functional modulations [165]. The full genome sequence is essential to provide knowledge for understanding natural genetic variations and development for breeding programs.

Recently, novel high-throughput sequencing strategies have revealed the structure of barley genome [166, 167]. Existence of 26,159 barley genes was confirmed by a systematic synteny analysis with model species from the Poaceae family (rice, maize, sorghum and *Brachypodium*) which have already had annotation of their genomes. Also, up to 80% of the 5.1 Gb genome of barley contains repetitive DNA, making the fully sequencing complicated [5]. Full annotations and a sequence-rich physical map of the barley genome, which is based on the genomic information contained in bacterial artificial chromosomes (BACs) developed for the Morex variety [168, 169], are available on public databases (http://webblast.ipk-gatersleben.de/barley/index.php) [167]. The first single nucleotide polymorphisms (SNPs) genotyping approach, based on the illumina oligo pool assays (OPAs), allowed the examination of 4596 markers in sets of 1536 SNPs [58]. Although declining cost of NGS technologies, thousands to million SNPs have been discovered via re-sequencing, providing greater detail for high density genetic maps [170].

Currently, array-based genotyping platform Infinium iSelect allows the simultaneous testing of 7842 SNPs [171]. Takahagi et al. [172], performed deep transcriptome sequencing, identified 38,729–79,949 SNPs in the 19 domesticated accessions and 55,403 SNPs in the wild barley. However, the complete sequences of the 525,599 bp mitochondrial genomes of wild and cultivated barley have been determined by Hisano et al. [135]. The mitochondrial genome of barley consists of 33 protein-coding genes, three ribosomal RNAs, 16 transfer RNAs, 188 new ORFs, six major repeat sequences and several types of transposable elements. The mitochondrial genomes of these wild and cultivated barley lines have been found to be almost identical in terms of both nucleotide sequence and genome structure, only three SNPs detected between haplotypes [135].

Several techniques, including linkage (or QTL mapping) mapping, association mapping (GWAS) and high-throughput omic techniques, such as transcriptomics, ionomics, proteomics and metabolomics analysis, have been used to identify a single gene or multi-genes corresponding to gene regulation networks of development, flowering, vernalization and biotic and abiotic stress conditions [173]. Next generation sequencing approaches (e.g. RNA-Seq) were carried out within 5 years and enlarged our knowledge about gene regulation networks of stress conditions. Especially, RNA-seq approach has been widely utilized due to low background noise, high sensitivity and reproducibility, great dynamic range of expression and base-pair resolution for transcription profiling [174]. Transcriptomic analyses of more than 28 plant species have revealed thousands of genes that are differentially regulated under drought stress conditions [175]. During last few years, an increasing number of these genes have been characterized and their function under drought conditions has been shown by the analysis of loss-of-function mutants or over expressing lines. Most of these functional characterization studies have been performed in the model species *Arabidopsis thaliana* and in the grass *Oryza sativa*. However, production of desired drought-tolerant crop species has required the identification of orthologous genes in each species. Transcriptome and whole-genome sequencing of different plant species lead to identify orthologous genes across several model and crop species [176].

Transcriptome profiling of barley under low nitrogen (LN) conditions have been determined by using RNA-seq approach. 1469 differentially expressed genes were identified between tolerant and sensitive barley varieties under LN. Differences between tolerant and sensitive genotypes involved transporters, transcription factors, kinases, antioxidant stress and hormone signalling related genes. However, DEGs were classified in amino acid metabolism, starch and sucrose metabolism, secondary metabolism [177]. Up to today, transcription dynamic of hulless barley grain development was not well understood. Tang et al. [178] have conducted comparative transcriptome approach to investigate changes during grain development. 38 DEGs were determined co-modulated in two barley landraces with the differential seed starch synthesis traits. The results showed that these 38 DEGs encoded proteins such as alpha-amylase-related proteins, lipid-transfer protein, homeodomain leucine zipper (HD-Zip), Nuclear Factor-Y, subunit B (NFYBs), as well as MYB transcription factors. Also, they found that two genes Hvulgare_GLEAN_10012370 and *Hvulgare*_GLEAN_10021199 encoding SuSy, AGPase (*Hvulgare*_GLEAN_10033640 and *Hvulgare*_GLEAN_10056301), as well as SBE2b (*Hvulgare*_GLEAN_10018352) were significantly contributed to the regulatory mechanism during grain development in both genotypes.

Numerous numbers of studies have been performed to understand biotic and abiotic stress tolerance mechanisms. For this purpose, RNA-seq approach or microarray have a valuable potential to define stress mechanisms. One of the studies has been conducted by Tombuloglu et al. [179] to discover the properties underlying the boron tolerance mechanism. By using transcriptome-wide approach, 256,847 unigenes were generated and, 16 and 17% of the transcripts were found to be differentially regulated in root and leaf tissues, respectively, according to gene expression analysis. Most of these unigenes were found to be involved in cell wall, stress response, membrane, protein kinase and transporter mechanisms [179]. Also, physiological and biochemical analysis have provided valuable insights towards a novel integrated molecular mechanism of stress tolerance mechanisms in barley. A genome-wide transcriptome analysis was performed to identify the mechanisms of cadmium (Cd) tolerance in two barley genotypes with distinct Cd tolerance by using microarray approach. Microarray expression profiling revealed that novel genes may play important roles in Cd tolerance which were mainly via producing protectants such as catalase against reactive oxygen species, Cd compartmentalization (e.g. phytochelatin-synthase and vacuolar ATPase) and defence response and DNA replication (e.g. chitinase and histones) [180]. Another study to understand abiotic stress was the sequencing of young leaves RNAs of wild barley treated with salt (500 mM NaCl) at four different time intervals. Differential expression profiles have been classified into nine clusters by two-dimensional hierarchical clustering. The most important groups were assigned to 'response to external stimulus' and 'electron-carrier activity' which means that the highly expressed transcripts are involved in several biological processes, including electron transport and exchanger mechanisms, flavonoid biosynthesis and reactive oxygen species (ROS) scavenging, ethylene production, signalling network and protein refolding [181]. Hulless barley, also called naked barley, often suffered from drought stress during growth and development. Therefore, Zeng et al. [182] have investigated co-regulated mRNAs expression patterns under early well water, later water deficit and finally water recovery treatments, and to identify mRNAs specific to water limiting conditions. The results showed that 853 DEGs were determined and categorized into nine clusters. The up-regulated genes were found to be relevant to abiotic stress responses in abscisic acid (ABA) dependent and independent signalling pathway, including NCED, PYR/PYL/RCAR, SnRK2, ABF, MYB/MYC, AP2/ERF family, LEA and DHN under low relative soil moisture content (RSMC) level. However, the transcriptome analysis revealed that the most affected genes were related to tetrapyrrole binding, photosystem and photosynthetic membrane under drought stress conditions.

The proteomic approach also plays significant roles to understand alterations in the context of physiological and morphological responses to biotic and abiotic stresses in barley. Rollins et al. [183] have investigated the proteins differentially regulated in response to drought, high temperature or a combination of both treatments by using differential gel electrophoresis and mass spectrometry. The study showed that the drought treatment induced strong reductions of biomass and yield, but not causing significant alterations in photosynthetic performance and the proteome. In contrast, the heat treatment and the combination of heat and drought caused the reduction of photosynthetic performance and changes of the leaf proteome. 14 proteins among 99 protein spots were identified as a genotype-specific manner in response to heat treatment. The analysis indicated that the differentially regulated proteins were related to photosynthesis, detoxification, energy metabolism and protein biosynthesis. Barley, also, used to identify the quantitative proteome changes under different drought conditions by Vítámvás et al. [184]. They cultivated plants for 10 days under different drought conditions that the soil water content was held at 65, 35 and 30% of soil water capacity (SWC), respectively. The proteomic alterations of barley crowns grown under different drought conditions were determined utilizing two-dimensional difference gel electrophoresis (2D-DIGE). Analysis of 2D-DIGE revealed that 105 differentially abundant spots were detected between the controls and drought-treated plants. The identified proteins were classified into stress-associated proteins, amino acid metabolism, carbohydrate metabolism, as well as DNA and RNA regulation and processing.

5. Genome editing

Genome editing has recently emerged as a novel transgenic method to improve crop plants has great opportunities over conventional gene targeted techniques. The most important advantage of gene editing is the modification of the targeting specific genes *in situ*. Genome editing, uses 'programmable' nucleases such as zinc finger nucleases (ZFNs), TAL effectors nucleases (TALENs) or clustered regularly inter-spaced short palindromic repeat (CRISPR)-associated endonucleases, may also be used to introduce gene insertions, gene replacements, insertions or deletions at specific genomic locations [185]. These proteins have a recognition domain, is provided by the FokI domain in both ZFNs and TALENs, and Cas9 in CRISPR systems, can be engineered to target specific sequences. Genome editing is based on double-strand break (DSB) induction [186], and subsequent repaired by the cell's own non-homologous end-joining (NHEJ) or homologous recombination (HR) mechanisms [185]. Genome editing is a key tool for advancing knowledge of gene function as well as allowing targeted mutagenesis with high efficiency in plants, including barley [187, 188].

Wendt et al. [187] reported the assembly of several TALENs for a specific genomic locus in barley. They tested the cleavage activity of individual TALENs *in vivo* using a yeast-based, single-strand annealing assay, and then the most efficient TALEN have been selected for barley transformation. Cleavage of the non-specific target was not observed, but analysis of the resulting transformants demonstrated that TALEN-induced double strand breaks led to the introduction of short deletions at the target site. Another study with TALENs has been reported by Gurushidze et al. [188] that they used TALENs in pollen-derived, regenerable cells to establish the generation of instantly true-breeding mutant plants. A gfp-specific TALEN pair was expressed via *Agrobacterium*-mediated transformation in embryogenic pollen with 22% of the TALEN transgenics. During gene replacement, desired DNA could integrate into the genome by homologous recombination that provides great promise to the introduction of mutations at pre-determined positions in the genome. Watanabe et al. [189] used a model system based on double-strand break induction by the mega nuclease I-SceI to target specific position in the genome. They obtained two transformants that were stably inherited as a single Mendelian trait. They suggested that stable gene replacement could be achieved in barley for routine applications by targeted double-strand break induction. The RNA-guided Cas9 system also represents a flexible approach for gene editing in barley and provides a valuable tool to create specific mutations that knock-out or alters target gene function. Lawrenson et al. [190] investigated the use and target specificity of RNA-guided Cas9 genome editing in barley. They demonstrated Cas9-induced mutations in the first generation of 23% for barley line. And also, they observed that stable Cas9-induced mutations were transmitted to T_2 plants independently of the T-DNA construct thus establishing the potential for rapid characterisation of gene function in barley.

6. Transposons, epigenetic studies and non-coding RNAs

Transposons, is a segment of DNA moves to new location in a chromosome or to another chromosome or cell, were first identified in maize by McClintock [191]. Several studies have been revealed that transposons affect gene structure, epigenetic regulations and genome dynamics of almost all living organisms [30]. Transposons alter the existing genome structure that can lead to significant changes such as deletions and/or insertions. Percentages and types of transposons can vary among species [192] that prokaryotic genomes contain 1–3% transposons. However, their percentage may reach 85% or more in eukaryotic genomes, especially plants [193]. Due to having larger genome, barley has larger transposon-derived DNA content with up to 85% [194]. Also, it was demonstrated that *Copia* retrotransposons remained intact and active for much longer time periods in the larger genomes such as barley than the smaller genomes [195].

Our group has been studying barley transposon effects on somaclonal variation, stability of aging barley calli and callus regeneration by using IRAP markers derived from *BARE-1* [47, 50] and *Nikita* [48]. In addition, mature embryo, leaf, root tissues were investigated for *BARE-1* and *BAGY2* movements by Marakli et al. [49] and *Sukkula* movement in barley, which is a non-autonomous retrotransposon, have been investigated by our group [51]. We demonstrated that *BAGY2* was more stable than *BARE-1*. Another study on transposon movements of retrotransposons and methylation alteration was performed by Temel and Gozukirmizi [196]. We found that not all callus induction conditions increased the retrotransposon activity. However, increase in cytosine methylation has been observed during callus formation using Sensitive Restriction Fingerprinting. Yilmaz et al. [197] also investigated the stability of aging barley calli and regenerated plantlets from those calli. We used the BAGY2 retrotransposon-specific IRAP technique to determine level of variations of DNA. We found that the culture conditions caused genetic variations, and also copy numbers of internal domains of *BAGY2* have increased. Moreover, IRAP technique has been utilized to assess the genotoxicity of some drugs such as epirubicin [52] and amiprophos-methyl [53]. Recently, Yuzbasioglu et al. [198] used IRAP markers to identify variation in single seed derived leaves and roots in rice.

Epigenetic chromatin modification is defined as heritable changes in gene expression which are not occurred by alterations in the nucleotide sequences of DNA. DNA methylation and modifications of covalent histone N-terminal tail are mainly regarded as chromatin modifications that can be changed in plants during the cell cycle [199, 200], plant development [201, 202] or in stress response [203]. The epigenetic mechanisms keep gene or genes active or repressive states [204, 205]. Braszewska-Zalewska and Hasterok [206] investigated the differences of epigenetic modification between root meristematic tissues of barley. Their study indicated that levels of epigenetic modifications varied between RAM tissues. Studies on environmental stresses showed that both DNA methylation and histone modifications are involved in DNA damage response. Also, Braszewska-Zalewska et al. [207] observed that chemical (maleic acid hydrazide; MH) and physical (gamma rays) mutagens strongly affected the level of histone methylation and acetylation. One of the major components of epigenetic variations is the combinations of histones carrying different covalent modifications that Baker et al. [208] have mapped nine modified histones in the barley seedling epigenome using chromatin immune precipitation next-generation sequencing (ChIP-seq) technique. They defined 10 chromatin states (five states to genes and five states to intergenic regions) representing local epigenetic environments in the barley genome. Moreover, it was found that H3K36me3-containing two genic states were related to constitutive gene expression. However, one genic state involving an H3K27me3 was related to differentially expressed genes.

The recent wide applications of whole-genome tilling array and RNA-sequencing (RNA-seq) approaches have revealed that the transcription landscape in eukaryotes is much more complex than had been expected [209]. These approaches have facilitated the identification of thousands of novel ncRNAs (or npcRNAs) in many organisms, such as humans, animals and plants [210–214]. ncRNAs are classified as short (<200 nt) and long ncRNAs (lncRNAs; >200 nt). Transcriptional and post-transcriptional regulation of gene expression of short ncRNAs, including siRNAs, miRNAs and piRNAs, has been well recognized and the molecular mechanisms of short ncRNA-mediated regulation have been well understood [215, 216]. On the contrary, the regulatory roles of lncRNAs are only beginning to be recognized and the molecular basis of lncRNA-mediated gene regulation is still poorly understood [217]. Our group has been investigating the association between salinity stress metabolism and barley lncRNAs (unpublished data). Identification of novel lncRNAs is likely to provide new insight into the complicated gene regulatory network involving lncRNAs, provide novel diagnostic opportunities, and pinpoint novel therapeutically targets.

7. Conclusion

Barley is an economically important crop plant, the fourth cereal worldwide in terms of the planting area, utilized almost 60% as animal feed, around 30% for malt production, 7% for seed production and only 3% for human food [11, 12]. In recent years, the malt derived from the germinated barley is the key material for the malting represents the most economically favourable application for beer brewing [13]. There is tremendous genetic research on barley at morphological, biochemical and molecular level for development of superior barley varieties. However, detailed analyses should be performed to investigate for the environmental extrapolation of laboratory developed lines. The relationship between environmental effects and genetic studies, especially field studies will provide knowledge about interaction of environment and genetically developed varieties. We tried to cite as many papers as possible. Yet we apologize to authors whose works are gone unmentioned in this chapter.

Acknowledgments

We are grateful to the Research Fund of Istanbul University for financial support (Projects FDK-2016-23086)

Author details

Nermin Gozukirmizi[1]* and Elif Karlik[2]

*Address all correspondence to: nermin@istanbul.edu.tr

1 Department of Molecular Biology and Genetics, Faculty of Science, Istanbul University, Istanbul, Turkey

2 Department of Biotechnology, Institution of Science, Istanbul University, Istanbul, Turkey

References

[1] Gujral HS, Gaur S. Instrumental texture of chapati as affected by barley flour, glycerol monostearate and sodium chloride. International Journal of Food Properties. 2005;8:377-385

[2] Shewry PR. Barley: genetics, biochemistry, molecular biology and biotechnology. Wallingford: C.A.B International; 1992. 085198-725-7.

[3] USDA, NRCS. The PLANTS database. National Plant Data Team, Greensboro, NC 2016; USA. NC27401-4901. http://plants.usda.gov, 17 November 2016

[4] von Bothmer R. The wild species of hordeum: Relationships and potential use for improvement of cultivated barley. In: SWEWRY PR, editor. Barley: Genetics, Biochemistry, Molecular Biology and Biotechnology. Willingford, Oxon: C.A.B International; 1992. pp. 3-18. Chapter 1

[5] The International Barley Genome Sequencing Consortium. A physical, genetic and functional sequence assembly of the barley genome. Nature. 2012;**491**:711-716

[6] Salamini F, Özkan H, Brandolini A, Schâfer-Pregl R, Martin W. Genetics and geography of wild cereal domestication in the near east. Nature Reviews Genetics. 2002;**3**(6): 429-441

[7] Harlan JR. Barley Hordeum vulgare (Gramineae--Triticinae). In: Simmonds NW, editor. Evolution in Crop Plants. London: Longman; 1976. pp. 93-98

[8] Poehlman JM. Adaptation and distribution. In: Rasmusson DC, editor. Barley. Madison, Wisconsin: American Society of Agronomy, Inc., Crop Science Society of America, Inc., Soil Science Society of America, Inc.; 1985. pp. 1-17

[9] Dunwell JM. Barley. In: Evans DA, Sharp R, Ammirato PV, editors. Handbook of Plant Cell Culture. Vol. 4. 1st ed. London: MacMillan and Co.; 1986. pp. 339-369

[10] Zhang J, Zheng HG, Aarti A, Pantuwan G, Nguyen TT, Tripathy JN, Sarial AK, Robin S, Babu RC, Nguyen BD, Sarkarung S, Blum A, Nguyen HT. Location genomic regions

associated with components of drought resistance in rice: Comparative mapping within and across species. Theoretical and Applied Genetics. 2001;**103**:19-29

[11] Baik BK, Ullrich SE. Barley for food: Characteristics, improvement, and renewed interest. Journal of Cereal Science. 2008;**48**:233-242

[12] Dawson IK, Russell J, Powell W, Steffenson B, Thomas WT, Waugh R. Barley: a translational model for adaptation to climate change. New Phytologist. 2015;**206**:913-31

[13] Bond J, Capehart T, Allen E, Kim G. Boutique brews, barley, and the balance sheet: Changes in malt barley industrial use require an updated forecasting approach. Economic Research Division, United Stated Department of Agriculture. 2015;18-23

[14] Lan W,Wang W, Yu Z, Qin Y, Luan J, Li X. Enhanced germination of barley (Hordeum vulgare L.) using chitooligosaccharide as an elicitor in seed priming to improve malt quality. Biotechnology Letters. 2016;**38**:1935-1940

[15] Holtekjolen AK, Baevere AB, Rodbotten M, Berg H, Knutsen SH. Antioxidant properties and sensory profiles of breads containing barley flour. Food Chemistry. 2008;**110**(2): 414-421

[16] Baba WN, Rashid I, Shah A, Ahmad M, Gani A, Masoodi FA. Effect of microwave roasting on antioxidant and anti-cancerous activities of barley flour. Journal Saudi Society of Agricultural Sciences. 2014;**27**:143-154

[17] Ahmad M, Gani A, Shah A, Gani A, Masoodi FA. Germination and microwave processing of barley (Hordeum vulgare L.) changes the structural and physicochemical properties of β-D-glucan & enhances its antioxidant potential. Carbohydrate Polymers. 2016;**153**:696-702

[18] Nevo E. Origin, evolution, population genetics and resources for breeding of wild barley, Hordeum spontaneum, in the Fertile Crescent. In: Shewry P, editor. Barley: Genetics, Molecular Biology and Biotechnology, C.A.B. International; Wallingford, U.K., 1992. pp. 19-43

[19] Lundqvist U. Scandinavian mutation research in barley – a historical review. Hereditas. 2014;**151**:123-131

[20] Technical Brochure (1970). Milne Marsters Company.

[21] Bilge E, Oraler G, Gozukirmizi N, Olgun A, Topaktafl M, Experimental mutations in barley (Hordeum vulgare L.). Istanbul Universitesi Fen Fakultesi Mec. 1981;**46**:29-35

[22] Bilge E, Oraler G, Gozukirmizi N, Olgun A, Topaktafl M, Cytogenetic studies on Hordeum vulgare L. treated with mutagens. Istanbul Universitesi Fen Fakultesi Mec. 1981;**46**:37-42

[23] Moose SP, Mumm RH. Molecular plant breeding as the foundation for 21 century crop improvement. Plant Physiology. 2008;**147**:969-977

[24] Rajib R, Abdelmoumen T, Hakeem KR, Mohamed RAG, Tah J. Molecular marker-assisted technologies for crop improvement. In: Roychowdhury R, editor. Crop improvement in

the era of climate change. Delhi, India: I.K. International Publication House Pvt. Ltd; 2013. pp. 241-258

[25] Foolad MR, Sharma A. Molecular markers as selection tools in tomato breeding. Acta Horticulturae. 2005;**695**:225-240

[26] Heun M, SchäFer-Pregl R, Klawan D, Castagna R, Accerbi M, Borghi B, Salamini F. Site of einkorn wheat domestication identified by DNA fingerprinting. Science. 1997;**278**:1312-1314

[27] Badr A, Müller K, Schäfer-Pregl R, El Rabey H, Effgen S, Ibrahim H, Pozzi C, Rohde W, Salamini F. On the origin and domestication history of barley (Hordeum vulgare). Molecular Biology and Evolution. 2000;**17**:499-510

[28] El Rabey HA, Badr A, Schäfer-Pregl R, Martin B, Salamini F. Species separation and incipient speciation in Hordeum (Poaceae) resolved by discontinuous molecular markers. Plant Biology. 2002;**4**:1-9

[29] El Rabey H, Khan JA, Al-Malki AL, Hussein HK. Synteny (co-linearity) in some cereal crops genomes as revealed by amplified fragment length polymorphisms (AFLP), simple sequence repeats (SSR) and inter simple sequence repeats (ISSR) markers. African Journal of Biotechnology. 2012;**11**(88):15387-15397

[30] Gozukirmizi N, Temel A, Maraklı S, Yilmaz S. Transposon activity in plant genomes. In: Hakeem KR, Tombuloglu H, Tombuloglu, editors. Plant Omics: Trends and Applications. London/Berlin, Switzerland: Springer; 2016. pp. 83-108

[31] Gupta PK, Rustgi S, Mir RR. Array-based high-throughput DNA markers for crop improvement. Heredity (Edinb) 2008;101:5-18

[32] Potokina E, Druka A, Luo Z, Wise R, Waugh R, Kearsey M. Gene expression quantitative trait locus analysis of 16000 barley genes reveals a complex pattern of genome wide transcriptional regulation. Plant Journal. 2008;**53**:90-101

[33] Springer NM, Ying K, Fu Y, Ji TJ, Yeh C-T, Jia Y, Wu W, Richmond TA, Kitzman J, Rosenbaum H, Iniguez AL, Barbazuk WB, Jeddeloh JA, Nettleton D, Schnable P. Maize inbreds exhibit high levels of copy number variation (CNV) and presence/absence variation (PAV) in genome content. PloS Genetics. 2009;**5**:11

[34] Belo A, Beatty MK, Hondred D, Fengler KA, Li B, Rafalski A. Allelic genome structural variations in maize detected by array comparative genome hybridization. Theoretical and Applied Genetics. 2010;**120**:355-367

[35] Barabaschi D, Tondelli A, Desiderio F, Volante A, Vaccino A, Valè G, Cattivelli L. Next generation breeding. Plant Science. 2016;**242**:3-13

[36] Tao F, Rötter R, Palosuo T, Díaz-Ambrona CGH, Mínguez-Tudela MI, Semenov M, Kersebaum KC, Nendel C, Cammarano D, Hoffmann H, Ewert F, Dambreville A, Martre P, Rodríguez L, Ruiz-Ramos M, Gaiser T, Höhn JG, Salo T, Ferrise R, Bindi M, Schulman AH. Designing future barley ideotypes using a crop model ensemble. European Journal Agronomy. 2016;**82**:144-162

[37] Graner A, Jahoor A, Schondelmaier J, Siedler H, Pillen K, Fischbeck G, Wenzel G, Herrmann RG. Construction of an RFLP map of barley. Theoretical and Applied Genetics, 1991;**83**(2):250-256

[38] Stein N, Prasad M, Scholz U, Thiel T, Zhang H, Wolf M, Kota R, Varshney RK, Perovic D, Grosse I, Graner A. A 1,000-loci transcript map of the barley genome: New anchoring points for integrative grass genomics. Theoretical and Applied Genetics. 2007;**114**(5):823-839

[39] Okumus A, Uzun F. Genetic and geographic variation of bulbous barley (Hordeum bulbosum L.) assessed by RAPD markers. Russian Journal of Genetics. 2007;**43**(3):294-298

[40] Agrama HA, Dahleen L, Wentz M, Jin Y, Steffenson B. Molecular mapping of the crown rust resistance gene Rpc1 in barley. Phytopathology. 2004;**94**(8):858-861

[41] Altinkut A, Kazan K, Gozukirmizi N. AFLP marker linked to water-stress-tolerant bulks in barley (Hordeum vulgare L.). Genetics and Molecular Biology. 2003;**26**(1):77-82

[42] Janiak A, Szarejko I. Molecular mapping of genes involved in root hair formation in barley. Euphytica. 2007;**157**(1-2):95-111

[43] Li XL, Yu XM, Wang NN, Feng QZ, Dong ZY, Liu LX, Shen JL, Liu B. Genetic and epigenetic instabilities induced by tissue culture in wild barley (Hordeum brevisubulatum (Trin.) Link). Plant, Cell, Tissue and Organ Culture. 2007;**90**(2):153-168

[44] Ablett GA, Karakousis A, Banbury L, Cakir M, Holton TA, Langridge P, Henry RJ. Application of SSR markers in the construction of Australian barley genetic maps. Australian Journal of Agricultural Research. 2003;**54**:1187-1195

[45] Karakousis A, Gustafson JP, Chalmers KJ, Barr AR, Langridge P. A consensus map of barley integrating SSR, RFLP, and AFLP markers. Australian Journal of Agricultural Research. 2003;**54**:1173-1185

[46] Tyrka M, Perovic D, Wardynska A, Ordon F. A new diagnostic SSR marker for selection of the Rym4/Rym5 locus in barley breeding. Journal of Applied Genetics. 2008;**49**(2):127-134

[47] Evrensel C, Yilmaz S, Temel A, Gozukirmizi N. Variations in BARE-1 insertion patterns in barley callus cultures. Genetics Molecular Research. 2011;**10**:980-987

[48] Bayram E, Yilmaz S, Hamat-Mecbur H, Kartal-Alacam G, Gozukirmizi N. Nikita retrotransposon movements in callus cultures of barley (Hordeumvulgare L). Plant Omics. 2012;**5**:211-215

[49] Marakli S, Yilmaz S, Gozukirmizi N. BARE1 and BAGY2 retrotransposon movements and expression analyses in developing barley seedlings. Biotechnology, Biotechnology Equipment. 2012;**26**:3451-3456

[50] Yilmaz S, Gozukirmizi N. Variation of retrotransposon movement in callus culture and regenerated shoots of barley. Biotechnology, Biotechnology Equipment. 2013;**27**:4227-4230

[51] Kartal-Alacam G, Yilmaz S, Marakli S, Gozukirmizi N. Sukkula retrotransposon insertion polymorphism in barley. Russian Journal of Plant Physiology. 2014;**61**:828-833

[52] Hamat-Mecbur H, Yilmaz S, Temel A, Sahin K, Gozukirmizi N. Effects of epirubicin on barley seedlings. Toxicology and Industrial Health. 2014;**30**:52-59

[53] Temel A, Gozukirmizi N. Genotoxicity of metaphase-arresting methods in barley. Turkish Journal of Biology. 2015;**39**:139-146. DOI: 10.3906/biy-1405-58.

[54] Russell JR, Dawson IK, Flavell A, Steffenson B, Weltzien E, Booth A, Ceccarelli S, Waugh R. Analysis of >1000 single nucleotide polymorphisms in geographically matched samples of landrace and wild barley indicates secondary contact and chromosome-level differences in diversity around domestication genes. New Phytologistogist. 2011;**191**: 564-578

[55] Muñoz-Amatriaín M, Eichten SR, Wicker T, Richmond TA, Mascher M, Steuernagel B, Scholz U, Ariyadasa R, Spannagl M, Nussbaumer T, Mayer KF, Taudien S, Platzer M, Jeddeloh JA,Springer NM, Muehlbauer GJ, Stein N. Distribution, functional impact, and origin mechanisms of copy number variation in the barley genome. Genome Biology. 2013;**14**(6): 58

[56] Albayrak G, Gozukirmizi N. RAPD Analysis of Genetic Variation in Barley. Turkish Journal of Agriculture and Forestry. 1999;**23**:627-630

[57] Mylonas IG, Georgiadis A, Apostolidis AP, Bladenopoulos K, Koutsika-Sotiriou M. Barley cultivar discrimination and hybrid purity control using RAPD markers. Romanian Biotechnological Letters. 2014;**19**(3):9421-9428

[58] Close TJ, Bhat PR, Lonardi S, Wu Y, Rostoks N, Ramsay L, et al. Development and implementation of high-throughput SNP genotyping in barley. BMCGenomics. 2009;10:582.

[59] Shavrukov Y. Comparison of SNP and CAPS markers application in genetic research in wheat and barley. BMC Plant Biology. 2016;**16**(1):11

[60] Zhang X, Lu C, Xu R, Zhou M. Development of molecular markers linked to barley heterosis. Euphytica. 2015;**203**:309-319

[61] Ovesná J, Kučera L, Vaculová K, Milotová J, Snape J, Wenzl P, et al. Analysis of the genetic structure of a barley collection using DNA Diversity Array Technology (DArT). Plant Molecular Biology Reporter. 2013;**31**:280-288

[62] Lex J, Ahlemeyer J, Friedt W, Ordon F. Genome-wide association studies of agronomic and quality traits in a set of German winter barley (Hordeum vulgare L.) cultivars using Diversity Arrays Technology (DArT). Journal of Applied Genetics. 2014;**55**:295-305

[63] Zhou H, Liu S, Liu Y, Liu Y, You J, Deng M, Ma J, Chen G, Wei Y, Liu C, Zheng Y.Mapping and validation of major quantitative trait loci for kernel length in wild barley (Hordeum vulgare ssp spontaneum). BMC Genetics. 2016;17(1):130

[64] Jia Q, Tan C, Wang J, Zhang XQ, Zhu J, Luo H, Yang J, Westcott S, Broughton S, Moody D, Li C. Marker development using SLAF-seq and whole-genome shotgun strategy to fine map the semi-dwarf gene ari-e in barley. BMC Genomics. 2016;**17**(1):911

[65] Ren X, Wang J, Liu L, Sun G, Li C, Luo H, Sun D. SNP-based high density genetic map and mapping of btwd1 dwarfing gene in barley. Scientific Reports. 2016;**6**:31741. PMID 27530597

[66] Zhou G, Zhang Q, Tan C, Zhang X, Li C. Development of genome-wide InDel markers and their integration with SSR, DArT and SNP markers in single barley map. BMC Genomics. 2015;**16**:804

[67] George EF. Plant Tissue Culture Procedure – Background, In: E.F. George, M.A. Hall, G.J. De Klerk (Eds.). Plant Propogation by Tissue Culture. 3rd ed. Dordrecht, the Netherlands: Springer; 2008. pp. 1-28

[68] Cardoza V. Tissue culture: The manipulation of plant development. In: Stewart Jr CN, editor. Plant Biotechnology and Genetics: Principles, Techniques and Applications. 1st ed. Hoboken, New Jersey: John Wiley and Sons; 2008. pp. 112-134

[69] Dahleen LS. Donor-plant environment effects on regeneration from barley embryo-derived callus. Crop Science. 1999;**39**(3):682-685

[70] Dahleen LS, Bregitzer P. An Improved Media System for High regeneration rates from barley immature embryo-derived callus cultures of commercial cultivars. Crop Science. 2002;**42**:934-938

[71] Sharma VK, Hansch R, Mendel RR, Schulze J. Mature embryo axis-based high frequency somatic embryogenesis and plant regeneration from multiple cultivars of barley (Hordeum vulgare L.). Journal of Experimental Botany. 2005;**56**(417):1913-1922

[72] Temel A, Gözükırmızı N. Advances in barley biotechnology: Tissue culture and molecular markers. In: Elfson SB, editor. Barley: Production, Cultivation and Uses. Ottawa: Nova Science Publishers; 2011. pp. 129-159

[73] Breiman A. Plant regeneration from hordeum spontaneum and hordeum bulbosum immature embryo derived calli. Plant Cell Reports. 1985;**4**(2):70-73

[74] Thomas MR, Scott KJ. Plant regeneration by somatic embryogenesis from callus initiated from immature embryos and immature inflorescences of Hordeum vulgare. Journal of Plant Physiology. 1985;**121**(2):159-169

[75] Goldstein CS, Kronstad WE. Tissue culture and plant regeneration from immature embryo explants of barley, Hordeum vulgare. Theoretical and Applied Genetics. 1986;**71**(4):631-636

[76] Jorgensen RB, Jensen CJ, Andersen B, Bothmer R. High capacity of plant regeneration from callus of interspecific hybrids with cultivated barley (Hordeum vulgare L.). Plant Cell, Tissue and Organ Culture. 1986;**6**(3):199-207

[77] Karp A, Steele SH, Breiman A, Shewry PRS, Parmar S, Jones MGK. Minimal variation in barley plants regenerated from cultured immature embryos. Genome. 1987;**29**:405-412

[78] Luhrs R, Lorz H. Plant regeneration in vitro from embryogenic cultures of spring- and winter-type barley (Hordeum vulgare L.) varieties. Theoretical and Applied Genetics. 1987;**75**(1):16-25

[79] Rotem-Abarbanell D, Breiman A. Plant regeneration from immature and mature embryo derived calli of Hordeum marinum. Plant Cell, Tissue and Organ Culture. 1989;**16**(3):207-216

[80] Rikiishi K, Matsuura T, Maekawa M, Takeda K. Light control of shoot regeneration in callus cultures derived from barley (Hordeum vulgare L.) immature embryos. Breeding Science. 2008;**58**(2):129-135

[81] Temel A, Kartal G, Gozukirmizi N. Genetic and epigenetic variations in barley calli cultures. Biotechnology and Biotechnological Equipment. 2008;**22**(4):911-914

[82] Lupotto E. Callus induction and plant regeneration from barley mature embryos. Annals of Botany. 1984;**54**:523-529

[83] Katoh Y, Hasegawa T, Suzuki T, Fujii T. Plant regeneration from the callus derived from mature embryos of hiproly barley, Hordeum distichum L. Agricultural and Biological Chemistry. 1986;**50**(3):761-762

[84] Ahloowalia BS. Plant regeneration from embryo-callus culture in barley. Euphytica. 1987;**36**(2):659-665

[85] Rengel Z. Embryogenic callus induction and plant regeneration from cultured Hordeum vulgare mature embryos. Plant Physiology and Biochemistry. 1987;**25**:43-48

[86] Ukai Y, Nishimura S. Regeneration of plants from calli derived from seeds and mature embryos in barley. Japanese Journal of Breeding. 1987;**37**:405-411

[87] Gozukirmizi N, Ari S, Oraler G, Okatan Y, Palavan N. Callus induction, plant regeneration and chromosomal variations in barley. Acta Botanica Neerlandica. 1990;**39**(4):379-387

[88] Cheng TY, Smith HH. Organogenesis from callus culture of Hordeum vulgare. Planta. 1975;**123**(3):307-310

[89] Weigel RC, Hughes KW. Long term regeneration by somatic embryogenesis in barley (Hordeum vulgare L.) tissue cultures derived from apical meristem explants. Plant Cell, Tissue and Organ Culture. 1985;**5**(2):151-162

[90] Ganeshan S, Baga M, Harwey BL, Rossnagel BG, Scoles GJ, Chibbar RN. Production of multiple shoots from thiadiazuron-treated mature embryos and leaf-base/apical meristems of barley (Hordeum vulgare L.). Plant Cell Tissue and Organ Culture. 2003;**73**:57-64

[91] Huang B, Sunderland N. Temperature-stress pretreatment in barley anther culture. Annals of Botany. 1982;**49**(1):77-88

[92] Piccirilli M, Arcioni S. Haploid plants regenerated via anther culture in wild barley (Hordeum spontaneum Kock C). Plant Cell Reports. 1991;**10**(6-7):273-276

[93] Hoekstra S, van Zijderveld MH, Louwerse JD, Heidekamp F, van der Mark F. Anther and microspore culture of Hordeum vulgare L. cv. Igri. Plant Science. 1992;**86**(1):89-96

[94] Bednarek PT, Orlowska R, Koebner RMD, Zimny J. Quantification of the tissue-culture induced variation in barley (Hordeum vulgare L.). BMC Plant Biology. 2007;**7**:10

[95] Kohler F, Wenzel G. Regeneration of isolated barley microspores in conditioned media and trials to characterize the responsible factor. Journal of Plant Physiology. 1985;**121**(2):181-191

[96] Kao KN, Saleem M, Abrams S, Pedras M, Horn D, Mallard C. Culture conditions for induction of green plants from barley microspores by anther culture methods. Plant Cell Reports. 1991;**9**(11):595-601

[97] Obert B, Middlefell-Williams J, Millam S. Genetic transformation of barley microspores using anther bombardment. Biotechnology Letters. 2008;**30**(5):945-949

[98] Castillo AM, Cistue L. Production of gynogenic haploids of Hordeum vulgare L. Plant Cell Reports. 1993;**12**(3):139-143

[99] Holme IB, Brinch-Pedersen H, Lange M, Holm PB. Transformation of different barley (Hordeum vulgare L.) cultivars by Agrobacterium tumefaciens infection of in vitro cultured ovules. Plant Cell Reports. 2008;**27**(12):1833-1840

[100] Kott LS, Kasha KJ. Initiation and morphological development of somatic embryoids from barley cell cultures. Canadian Journal of Botany. 1984;62(6):1245-1249.

[101] Seguin-Swartz G, Kott L, Kasha KJ. Development of haploid cell lines from immature barley, Hordeum vulgare, embryos. Plant Cell Reports. 1984;**3**(3):95-97

[102] Luhrs R, Lorz H. Initiation of morphogenic cell-suspension and protoplast cultures of barley (Hordeum vulgare L.). Planta. 1988;**175**(1):71-81

[103] Muller B, Schulze J, Wegner U. Establishment of barley cell suspension cultures of mesocotyl origin suitable for isolation of dividing protoplasts. Biochemie und Physiologie der Pflanzen. 1989;**185**(1-2):123-130

[104] Luhrs R, Nielsen K. Microspore cultures as donor tissue for the initiation of embryogenic cell suspensions in barley. Plant Cell, Tissue and Organ Culture. 1992;**31**(2):169-178

[105] Lazzeri PA, Brettschneider R, Luhrs R, Lorz H. Stable transformation of barley via PEG-induced direct DNA uptake into protoplasts. Theoretical and Applied Genetics. 1991;**81**(4):437-444

[106] Sahrawat AK, Chand S. High frequency plant regeneration from coleoptile tissue of barley (Hordeum vulgare L.). Plant Science. 2004;**167**(1):27-34

[107] Li HP, Huang T, Wang CX, Liao YC. An efficient regeneration system of barley cultivars from leaf base segments. Biologia Plantarum. 2009;**53**(4):733-736

[108] Vitanova Z, Vitanov V, Trifonova A, Savova D, Atanasov A. Effect of 2, 4-D precultivation on regeneration capacity of cultivated barley. Plant Cell Reports. 1995;**14**:437-441

[109] Chang Y, Zitzewitz J, Hayes PM, Chen THH. High frequency plant regeneration form immature embryos of elite barley cultivars (Horedum vulgare L cv. Morex). Plant Cell Reports. 2003;**21**:733-738

[110] Bregitzer P. Plant regeneration and callus type in barley: Effect of genotype and culture medium. Crop Science. 1992;**32**:1108-1112

[111] Castillo AM, Egana B, Sanz JM, Cistue L. Somatic embryogenesis and plant regeneration from barley cultivars grown in Spain. Plant Cell Repots. 1998;**17**:902-906

[112] Sener O, Can E, Arslan M, Çeliş N. Effects of genotype and picloram concentrations on callus induction and plant regeneration from immature inflorescence of spring barley cultivars (Hordeum vulgare L.). Biotechnology & Biotechnological Equipment. 2008;**22**:915-920

[113] He T, Jia JF. High frequency plant regeneration from mature embryo explants of highland barley (Hordeum vulgare L. var. nudum Hk. f.) under endosperm-supported culture. Plant Cell Tissue and Organ Culture. 2008;**95**(2):251-254

[114] Bregitzer P, Campbell RD. Genetic markers associated with green and albino plant regeneration from embryogenic barley callus. Crop Science. 2001;**41**:173-179

[115] Gaponenko AK, Petrova TF, Iskakov AR, Sozinov AA. Cytogenetics of in vitro cultured somatic cells and regenerated plants of barley (Hordeum vulgare L.). Theoretical and Applied Genetics. 1998;**75**(6):905-911

[116] Choi HW, Lemaux PG, Cho MJ. High frequency of cytogenetic aberration in transgenic oat (Avena sativa L.) plants. Plant Science. 2000;**160**(4):761-772

[117] Choi HW, Lemaux PG, Cho MJ. Increased chromosomal variation in transgenic versus nontransgenic barley (Hordeum vulgare L.) plants. Crop Science. 2000;**40**:524-533

[118] Bregitzer P, Dahleen LS, Neate S, Schwarz P, Mancharan M. A single backcross effectively eliminates agronomic and quality alterations caused by somaclonal variation in transgenic barley. Crop Science. 2008;**48**(2):471-479

[119] Jiang W, Cho MJ, Lemaux PG. Improved callus quality and prolonged regenerability in model and recalcitrant barley (Hordeum vulgare L.) cultivars. Plant Biotechnology. 1998;**15**:63-69

[120] Espinasse A, Lay C. Shoot regeneration of callus derived from globular to torpedo embryos from 59 sunflower genotypes. Crop Science. 1989;**29**:201-205

[121] Bregitzer P, Campbell RD, Wu Y. Plant regeneration from barley callus: effects of 2, 4-dichlorophenoxyacetic acid and ohenylacetic acid. Plant Cell, Tissue and Organ Culture. 1995;**43**:229-235

[122] Chauhan M, Kothari SL. Optimization of nutrient levels in the medium increase the effeciency of callus induction and plant regeneration in recalcitrant Indian barley (Hordeum vulgare L.) in vitro. In Vitro Cellular and Developmental Biology-Plant. 2004;**40**:520-527

[123] Šerhantová V, Ehrenbergerová J, Ohnoutková L. Callus induction and regeneration efficiency of spring barley cultivars registered in the Czech Republic. Plant, Soil and Environment. 2004;**50**:456-462

[124] Nasircilara G, Kenan T, Callan F. Callus induction and plant regeneration from mature embryos of different wheat genotypes. Pakistan Journal of Botany. 2006;**38**(2):637-645

[125] Ward KA, Jordan MC. Callus formation and plant regeneration from immature and mature embryos of rye (Secale cereale L.). In vitro Cellular and Developmental Biology – Plant. 2001;**37**:361-268

[126] Fahmy AH, El-Shihy O. Improvement of plant regeneration from long term callus cultures of two Egyptian wheat cultivars. Arab Journal of Biotechnology. 2006;**8**(1):177-188

[127] Ozawa K, Komamine A. Establishment of a system of high-frequency embryogenesis from long-term cell suspension cultures of rice (Oryza sativa L.). Theoretical Applied Genetic. 1989;**77**:205-211

[128] Denchev PO, Conger BV. Plant regeneration from callus cultures of switchgrass. Crop Science. 1994; 34:1623-1627.

[129] Przetakiewicz A, Orczyk W, Nadolska-Orczyk A. The effect of auxin on plant regeneration of wheat, barley and triticale. Plant Cell, Tissue Organ Culture. 2003;**73**:245-256

[130] Murray F, Brettell R, Matthews P, Bishop D, Jacobsen J. Comparison of Agrobacterium-mediated transformation of four barley cultivars using the GFP and Gus reporter genes. Plant Cell Reports. 2004;**22**:397-402

[131] Travella S, Ross SM, Harden J, Everett C, Snape JW, Harwood WA. A comparison of transgenic barley lines produced by particle bombardment and Agrobacterium-mediated techniques. Plant Cell Reports. 2005;**23**:780-789

[132] Hanzel JJ, Miller JP, Brinkman MA, Fendos E. Genotype and media effects on callus formation and regeneration in barley. Crop Science. 1985;**25**:27-31

[133] von Arnold S, Sabala I, Bozhkov P, Dyachok J, Filonova L. Development pathways of somatic embryogenesis. Plant Cell Tissue Organ Culture. 2002;**69**:233-249

[134] Marthe C, Kumlehn J, Hensel G. Barley (Hordeum vulgare L.) transformation using immature embryos. Methods in Molecular Biology. 2015;**1223**:71-83

[135] Hisano H, Matsuura T, Mori IC, Yamane M, Sato K. Endogenous hormone levels affect the regeneration ability of callus derived from different organs in barley. Plant Physiology and Biochemistry. 2016;**99**:66-72

[136] Powell W, Dunwell JM. In vitro genetics of barley (Hordeum vulgare L.) 1. Response of immature embryos to 2, 4-dichlorophenozyacetic acid. Heredity. 1987;**59**:293-299 DOI:10.1038/hdy.1987.126.

[137] Baillie AMR, Rossnagel BG, Kartha KK. Evaluation of 10 Canadian barley (Hordeum vulgare L.) cultivars for tissue culture response. Canadian Journal of Plant Science. 1993;**73**:171-174

[138] Tiidema A, Truve E. Efficient regeneration of fertile barley plants from callus cultures of several Nordic cultivars. Hereditas. 2004;**140**:171-176

[139] Gozukirmizi N, Arı S, Gürel F, Gümüsel F, Çirakoglu B. Fingerprinting barley genome using PCR with arbitrary primers in barley regenerated from tissue culture. Proceedings of the Asia Pacific Conference on Agricultural Biotechnology; 20-24 August 1992; Science and Technology Press, Beijing, China; 1992. pp.143-146

[140] Rostami H, Giri A, Nejad ASM, Moslem A. Optimization of multiple shoot induction and plant regeneration in Indian barley (Hordeum vulgare) cultivars using mature embryos. Saudi Journal of Biology Sciences. 2013;**20**(3):251-255

[141] Wojnarowiez G, Jacquard C, Devaux P, Sangwan RS, Clément C. Influence of copper sulfate on anther culture in barley (Hordeum vulgare L.). Plant Science. 2002;**162**:843-847

[142] Tingay S, McElroy D, Kala R, Fieg S, Wang M, Thornton S, Brettell R. Agrobacterium tumefaciens-mediated barley transformation. Plant Journal. 1997;**11**:1369-1376

[143] Matthews PR, Wang MB, Waterhouse PM, Thornton S, Fieg SJ, Gubler F Jacobsen JV. Marker gene elimination from transgenic barley, using co-transformation with adjacent `twin T-DNAs' on a standard Agrobacterium transformation vector. Molecular Breeding. 2001;**7**(3):195-202

[144] Trifonova A, Madsen S, Olesen A. Agrobacterium-mediated transgene delivery and integration into barley under a range of in vitro culture conditions. Plant Science. 2001;**161**(5):871-880

[145] Ashok KS, Becker D, Lorz H. Agrobacterium tumefaciens-mediated genetic transformation of barley (Hordeum vulgare L.). Plant Science. 2007;**172**:281-290

[146] Kumlehn J, Serazetdinova L, Hensel G, Becker D, Loerz H. Genetic transformation of barley (Hordeum vulgare L.) via infection of androgenetic pollen cultures with Agrobacterium tumefaciens. Plant Biotechnology Journal. 2006;**4**(2):251-261

[147] Gurel F, Gozukirmizi N. Optimization of gene transfer into barley (Hordeum vulgare L.) mature embryos by tissue electroporation. Plant Cell Reports. 2000;**19**:787-791

[148] Bartlett JG, Alves SC, Smedley M, Snape LW, Harwood WA. High-throughput Agrobacterium-mediated barley transformation. Plant Methods. 2008;**4**:22

[149] Hensel G, Valkov V, Middlefell-Williams J, Kumlehn J. Efficient generation of transgenic barley: The way forward to modulate plant–microbe interactions. Journal of Plant Physiology. 2008;**165**:71-82

[150] Harwood WA, Ross SM, Bulley SM, Travella S, Busch B, Harden J, Snape JW. Use of the firefly luciferase gene in a barley (Hordeum vulgare) transformation system. Plant Cell Reports. 2002;**21**:320-326

[151] Hisano H, Sato K. Genomic regions responsible for amenability to Agrobacterium-mediated transformation in barley. Scientific Reports. 2016;**6**:37505

[152] Gozukirmizi N, Temel A. Advances in barley biotechnology: Tissue culture and molecular markers. In: Elfson SB, editor. Barley: Production, Cultivation and Uses. Ottawa: Nova Science Publishers; 2011. pp. 129-159

[153] Yeo FKS, Hensel G, Vozábová T, Martin-Sanz A, Marcel TC, Kumlehn J, Niks RE. Golden SusPtrit: a genetically well transformable barley line for studies on the resistance to rust fungi. Theoretical and Applied Genetics. 2014;**127**:325

[154] Nagy B, Majer P, Mihály R, Pauk J, Horváth GV. Stress tolerance of transgenic barley accumulating the alfalfa aldose reductase in the cytoplasm and the chloroplast. Phytochemistry. 2016;**129**:14-23

[155] Han N, Na C, Chai Y, Chen J, Zhang Z, Bai B, Bian H, Zhang Y, Zhu M. Over-expression of (1,3;1,4)-β-D-glucanase isoenzymes EII gene results in decreased (1,3;1,4)-β-D-glucan content and increased starch level in barley grains. Journal of the Science of Food and Agriculture. 2017;**97**:122-127. DOI: 10.1002/jsfa.7695.

[156] Wan Y, Lemaux PG. Generation of large number of independently transformed fertile barley plants. Plant Physiology. 1994;**104**:37-48

[157] Ritala A, Aspegren K, Kurten U, Salmenkallio-Martilla M, Mannonen L, Hannus R, Kaupennin R, Terri H, Enari T. Fertile transgenic barley by particle bombardment of immature embryos. Plant Molecular Biology. 1994;**24**:317-325

[158] Cho M, Jiang B, Lemaux G. Transformation of recalcitrant barley cultivars through improvement of regenerability and decreased albinism. Plant Sciences. 1998;**138**:229-244

[159] Yadav T, Kachhwaha S, Kothari SL. Efficient in vitro plant regeneration and generation of transgenic plants in barley (Hordeum vulgare L.) using particle bombardment. Journal of Plant Biochemistry and Biotechnology. 2013;**22**(2):202-213

[160] Zhang Z, Xing A, Staswick P, Clemente T. The use of glufosinate as a selective agent in Agrobacterium-mediated transformation of soybean. Plant Cell Tissue and Organ Culture. 1999;**56**:37-46

[161] Holme IB, Brinch-Pedersen H, Lange M, Holm PB. Transformation of barley (Hordeum vulgare L.) by Agrobacterium tumefaciens infection of in vitro cultured ovules. Plant Cell Reports. 2006;**25**:1325-1335

[162] Kovalchuk N, Jia W, Eini O, Morran S, Pyvovarenko T, Fletcher S, Bazanova N, Harris J, Beck-Oldach K, Shavrukov Y, Langridge P, Lopato S. Optimization of TaDREB3 gene expression intransgenic barley using cold-inducible promoters. Plant Biotechnology Journal. 2013;**11**:659-670

[163] Gürel F, Uçarli C, Tufan F, Kalaskar DM. Enhancing T-DNA transfer efficiency in barley (Hordeum vulgare L.) cells using extracellular cellulose and lectin. Applied Biochemistry and Biotechnology. 2015;**176**:1203-1216

[164] Uçarli C, Tufan F, Gürel AF. Expression and genomic integration of transgenes after Agrobacterium-mediated transformation of mature barley embryos. Genetics and Molecular Research. 2015;**14**:1096-1105

[165] Barabaschi D, Guerra D, Lacrima K, Laino P, Michelotti V, Urso S, Valè G, Cattivelli L. Emerging Knowledge from genome sequencing of crop species. Molecular Biotechnology. 2011;**50**:250-266

[166] Mayer KFX, Martis M, Hedley PE, Simkova H, Liu H, Morris JA, et al. Unlocking the barley genome by chromosomal and comparative genomics. Plant Cell. 2011;**23**:1249

[167] Mayer KFX, Waugh R, Langridge P, Close TJ, Wise RP, Graner A, et al. A physical, genetic and functional sequence assembly of the barley genome. Nature. 2012;**491**:711

[168] Schulte D, Ariyadasa R, Shi B, Fleury D, Saski C, Atkins M, deJong P, Wu CC, Graner A, Langridge P, Stein N. BAC library resources for map-based cloning and physical map construction in barley (Hordeum vulgare L.). BMC Genomics. 2011;**12**:247

[169] Yu X, Yang J, Li XR, Liu XX, Sun CB, Wu FJ, He Y. Global analysis of cis-natural antisense transcripts and their heatresponsive nat-siRNAs in Brassica rapa. BMC Plant Biology. 2013;**13**:208

[170] Muñoz-Amatriaín M, Moscou MJ, Bhat PR, Svensson JT, Bartoš J, Suchánková P, et al. An improved consensus linkage map of barley based on flow sorted chromosomes and single nucleotide polymorphism markers. Plant Genome. 2011;**4**:238-49

[171] Muñoz-Amatriaín M, Cuesta-Marcos A, Endelman JB, Comadran J, Bonman JM, Bockelman HE, Chao S, Russell J, Waugh R, Hayes PM, Muehlbauer GJ. The USDA barley core collection: genetic diversity, population structure, and potential for genome-wide association studies. PLoS One. 2014;**9**(4):e94688

[172] Takahagi K, Uehara-Yamaguchi Y, Yoshida T, Sakurai T, Shinozaki K, Mochida K, Saisho D. Analysis of single nucleotide polymorphisms based on RNA sequencing data of diverse biogeographical accessions in barley. Scientific Reports. 2013;**6**:33199

[173] Urano K, Kurihara Y, Seki M, Shinozaki K. 'Omics' analyses of regulatory networks in plant abiotic stress responses. Current Opinion of Plant Biology. 2010;**13**:132-138

[174] Marioni JC, Mason CE, Mane SM, Stephens M, Gilad Y. RNA-Seq: An assessment of technical reproducibility and comparison with gene expression arrays. Genome Research. 2008;**18**(9):1509-1517

[175] Deyholos MK. Making the most of drought and salinity transcriptomics. Plant Cell Environment. 2010;**33**: 648-654

[176] Alter S, Bader KC, Spannagl M, Wang Y, Bauer E, Schön CC, Mayer KF. DroughtDB: An expert-curated compilation of plant drought stress genes and their homologs in nine species. Database (Oxford). 2015;**2015**: bav046. DOI: 10.1093/database/bav046.

[177] Quan X, Zeng J, Ye L, Chen G, Han Z, Shah JM, Zhang G. Transcriptome profiling analysis for two Tibetan wild barley genotypes in responses to low nitrogen. BMC Plant Biology. 2016;**16**:30

[178] Tang Y, Zeng X, Wang Y, Bai L, Xu Q, Wei Z, Yuan H, Nyima T. Transcriptomics analysis of hulless barley during grain development with a focus on starch biosynthesis. Function and Integrative Genomics. 2017;**17**(1):107-117

[179] Tombuloglu G, Tombuloglu H, Sakcali MS, Unver T. High-throughput transcriptome analysis of barley (Hordeum vulgare) exposed to excessive boron. Gene. 2015;**557**(1):71-81

[180] Cao F, Chen F, Sun H, Zhang G, Chen ZH, Wu F. Genome-wide transcriptome and functional analysis of two contrasting genotypes reveals key genes for cadmium tolerance in barley. BMC Genomics. 2014;**15**:611

[181] Bahieldin A, Atef A, Sabir JS, Gadalla NO, Edris S, Alzohairy AM, Radhwan NA, Baeshen MN, Ramadan AM, Eissa HF, Hassan SM, Baeshen NA, Abuzinadah O, Al-Kordy MA, El-Domyati FM, Jansen RK. RNA-Seq analysis of the wild barley (H. spontaneum) leaf transcriptome under salt stress. Compets Rendus Biologies. 2015;**338**(5):285-297

[182] Zeng X, Bai L, Wei Z, Hongjun Y, Wang Y, Xu Q, Tang Y, Nyima T. Transcriptome analysis revealed the drought-responsive genes in Tibetan hulless barley. BMC Genomics. 2016;**17**:386

[183] Rollins JA, Habte E, Templer SE, Colby T, Schmidt J, von Korff M. Leaf proteome alterations in the context of physiological and morphological responses to drought and heat stress in barley (Hordeum vulgare L.). Journal of Experimental Botany. 2013;**64**(11):3201-3212

[184] Vítámvás P, Urban MO, Škodáček Z, Kosová K, Pitelková I, Vítámvás J, Renaut J, Prášil IT. Quantitative analysis of proteome extracted from barley crowns grown under different drought conditions. Frontiers in Plant Science. 2015;**6**:479

[185] Sprink T, Metje J, Hartung F. Plant genome editing by novel tools: TALEN and other sequence specific nucleases. Current Opinion in Biotechnology. 2015;**32**:47-53

[186] Bogdanove AJ, Voytas DF. 2011. TAL Effectors: Customizable proteins for DNA targeting. Science. 2011;**333**(6051):1843-1846

[187] Wendt T, Holm PB, Starker CG, Christian M, Voytas DF, Brinch-Pedersen H, Holme IB. TAL effector nucleases induce mutations at a pre-selected location in the genome of primary barley transformants. Plant Molecular Biology. 2013;**83**(3):279-285

[188] Gurushidze M, Hensel G, Hiekel S, Schedel S, Valkov V, Kumlehn J. True-breeding targeted gene knock-out in barley using designer TALE-nuclease in haploid cells. PLoS ONE. 2014;**9**(3):e92046

[189] Watanabe K, Breier U, Hensel G, Kumlehn J, Schubert I, Reiss B. Stable gene replacement in barley by targeted double-strand break induction. Journal of Experimental Botany. 2016;**67**(5):1433-1445

[190] Lawrenson T, Shorinola O, Stacey N, Li C, Ostergaard L, Patron N, Uauy C, Harwood W. Induction of targeted, heritable mutations in barley and Brassica oleracea using RNA-guided Cas9 nuclease. Genome Biology. 2015;**16**:258

[191] McClintock B. The significance of responses of the genome to challenge. Science. 1984;**226**:792-801

[192] Feschotte C, Pritham EJ. DNA transposons and the evolution of eukaryotic genomes. Annual Review Genetics. 2007;**41**:331-368

[193] Cordaux R, Batzer MA. The impact of retrotransposons on human genome evolution. Nature Review Genetics. 2009;**10**:691-703

[194] Wicker T, Zimmermann W, Perovic D, Paterson AH, Ganal M, Graner A, Stein N. A detailed look at 7 million years of genome evolution in a 439-kb contiguous sequence at the barley Hv-eIF4E locus: recombination, rearrangements and repeats. Plant Journal. 2015;**41**:184-194

[195] Wicker T, Keller B. Genome-wide comparative analysis of copia retrotransposons in Triticeae, rice, and Arabidopsis reveals conserved ancient evolutionary lineages and distinct dynamics of individual copia families. Genome Research. 2007;**17**:1072-1081

[196] Temel A, Gozukirmizi N. Analysis of retrotransposition and DNA methylation in barley callus culture. Acta Biologica Hungarica. 2013;**64**:86-95

[197] Yilmaz S, Marakli S, Gozukirmizi N. BAGY2 retrotransposon analyses in barley calli cultures and regenerated plantlets. Biochemistry Genetics. 2014;**52**:233-244

[198] Yuzbasioglu G, Yilmaz S, Marakli S, Gozukirmizi N. Analysis of Hopi/Osr27 and Houba/Tos5/Osr13 retrotransposons in rice. Biotechnology & Biotechnological Equipment. 2016;**30**:213-218

[199] Jasencakova Z, Meister A, Schubert I. Chromatin organization and its relation to replication and histone acetylation during the cell cycle in barley. Chromosoma. 2001;**110**:83-92

[200] Jasencakova Z, Soppe WJ, Meister A, Gernand D, Turner BM, Schubert I. Histone modifications in Arabidopsis—high methylation of H3 lysine 9 is dispensable for constitutive heterochromatin. Plant Journal. 2003;**33**:471-480

[201] Santamaría ME, Hasbún R, Valera MJ, Meijón M, Valledor L, Rodríguez JL, Toorop PE, Cañal MJ, Rodríguez R. Acetylated H4 histone and genomic DNA methylation patterns during bud set and bud burst in Castanea sativa. Journal of Plant Physiology. 2009;**166**:1360-1369

[202] Meijón M, Feito I, Valledor L, Rodríguez R, Cañal MJ. Dynamics of DNA methylation and Histone H4 acetylation during floral bud differentiation in azalea. BMC Plant Biology. 2010;**10**:10

[203] Luo M, Liu X, Singh P, Cui Y, Zimmerli L, Wu K. Chromatin modifications and remodeling in plant abiotic stress responses. Biochimistry and Biophysics Acta. 2012;**1819**:129-136

[204] Finnegan EJ, Genger RK, Peacock WJ, Dennis ES. DNA methylation in plants. Annual Review of Plant Physiology and Plant Molecular Biology. 1998;**49**(1):223-247

[205] Cheung P, Lau P. Epigenetic regulation by histone methylation and histone variants. Molecular Endocrinology. 2005;**19**(3):563-573

[206] Braszewska-Zalewska A, Hasterok R. Epigenetic modifications of nuclei differ between root meristematic tissues of Hordeum vulgare. Plant Signal Behavior. 2013;**8**(10): e26711

[207] Braszewska-Zalewska A, Tylikowska M, Kwasniewska J, Szymanowska-Pulka J. Epigenetic chromatin modifications in barley after mutagenic treatment. Journal of Applied Genetics. 2014; **55**(4):449-456

[208] Baker K, Dhillon T, Colas I, Cook N, Milne I, Milne L, Bayer M, Flavell AJ. Chromatin state analysis of the barley epigenome reveals a higher-order structure defined by H3K27me1 and H3K27me3 abundance. Plant Journal. 2015;84(1):111-24

[209] Jacquier A. The complex eukaryotic transcriptome: Unexpected pervasive transcription and novel small RNAs. Nature Review Genetics. 2009;**10**:833-844

[210] Ravasi T, Suzuki H, Pang KC, Katayama S, Furuno M, Okunishi R, et al. Experimental validation of the regulated expression of large numbers of non-coding RNAs from the mouse genome. Genome Research. 2006;**16**:11-9

[211] Consortium EP, Birney E, Stamatoyannopoulos JA, Dutta A, Guigo R, Gingeras TR, et al. Identification and analysis of functional elements in 1% of the human genome by the ENCODE pilot project. Nature. 2007;**447**:799-816

[212] Guttman M, Amit I, Garber M, French C, Lin MF, Feldser D, et al. Chromatin signature reveals over a thousand highly conserved large non-coding RNAs in mammals. Nature. 2009;458:223-7

[213] Matera AG, Terns RM, Terns MP. Non-coding RNAs: Lessons from the small nuclear and small nucleolar RNAs. Nature Reviews Molecular Cell Biology. 2007;**8**:209-220

[214] Ponting CP, Oliver PL, Reik W. Evolution and functions of long noncoding RNAs. Cell. 2009;**136**:629-641

[215] Ghildiyal M, Zamore PD. Small silencing RNAs: An expanding universe. Nature Review Genetics. 2009;**10**:94-108

[216] Chitwood DH, Timmermans MC. Small RNAs are on the move. Nature. 2010;**467**:415-419

[217] Qian-Hao Z, Ming-Bo W. Molecular functions of long non-coding RNAs in plants. Genes. 2012;**3**(1):176-190

Section 2

Microbiology of Brewing

Chapter 4

Saccharomyces and Non-*Saccharomyces* Starter Yeasts

Marilena Budroni, Giacomo Zara,
Maurizio Ciani and Francesca Comitini

Additional information is available at the end of the chapter

http://dx.doi.org/10.5772/intechopen.68792

Abstract

This chapter describes the importance of yeast in beer fermentation. Initially, the differences between *Saccharomyces cerevisiae* and *Saccharomyces pastorianus* in the production of "ale" and "lager" beers are analyzed. Then, the relationships between beer nutrients and yeast growth are discussed, with special emphasis on the production of the flavor compounds. The impact of the wort composition on flocculation is also discussed. Furthermore, conventional approaches to starter yeast selection and the development of genetically modified microorganisms are analyzed. Recent discoveries relating to the use of *S. cerevisiae* strains isolated from different food matrices (i.e., bread and wine) and the potential for the use of non-*Saccharomyces* starter strains in beer production are discussed. A detailed review of the selection of starter strains for the production of specialty beers then follows, such as for gluten-free beers and biologically aged beers. Yeast recovery from top-cropping and bottom-cropping systems and the methodologies and issues in yeast propagation in the laboratory and brewery (i.e., re-pitching) are also analyzed. Finally, the available commercial preparations of starter yeast and the methods to evaluate yeast viability prior to inoculation of the must are analyzed.

Keywords: yeast, *Saccharomyces*, non-*Saccharomyces*, fermentation, aroma

1. Introduction

Beer is a very old biotechnology, with its origins dating back to around 10,000 years ago [1]. Due to this ancient history, this alcoholic beverage has undergone three particularly important revolutions: (i) the practice of the *inoculum* of a selected yeast culture; (ii) the craft and home brewing movement from Europe to USA; and (iii) the influence of the genomic era and "big data" that now allow comparative genomic analyses that have provided new knowledge about yeast. The interest in comparative genomics of the yeast genome is in part motivated by the rationale of "beer *du terroir*": beers produced with local raw materials and local yeast. In this context, yeast appear to be a relatively unexplored tool for the diversification of local beers, as much of the characteristics of such beers are due to the yeast strains used. Indeed, the transformation of wort into beer essentially represents the yeast-driven conversion of sugars into ethanol, CO_2, and many other secondary products that provide specific aromas and flavors.

All brewers know that during the fermentation of the wort, the yeast have to complete two major tasks. The primary task is, as indicated, to convert the sugar into ethanol, CO_2, and the aromas for the production of quality beer, while they also need growth to produce biomass that will be re-pitchable into new brews. Thus, it is imperative to satisfy the physical and nutritional requirements of the yeast. The growth rate of yeast can be modulated by manipulation of some of the fermentation parameters, such as the supply of nutrients, the dissolved oxygen, and the temperature. Furthermore, to end up with a high quality beer, there must be a balance between the nutrients absorbed and the products released.

Beer production is essentially a two-phase process: primary fermentation and secondary fermentation (or maturation). Primary fermentation is a short and vigorous step, during which almost all of the sugars are fermented, accompanied by the production of the secondary compounds that result from the yeast metabolism, most of which are associated with the final beer aroma. At the end of this stage, most of the yeast biomass is collected and separated from the "green" beer, which then undergoes secondary fermentation. During this second phase, the yeast completes the fermentation of the residual sugar, with undesirable compounds removed, and the final taste of beer defined. Among the fermentative yeast, *Saccharomyces cerevisiae* and *Saccharomyces pastorianus* are the most frequently used in starter cultures in breweries. As the species of the genus *Saccharomyces*—and particularly *S. cerevisiae*—dominated during spontaneous fermentation, they were selected over the years across different cultures and from distinct environments. In particular, in the beer industry, the goal of the use of inoculated yeast is to increase the fermentation efficiency, to develop new beers, and especially to enhance the sensory complexity of the beer that is produced.

2. Yeast in beer

2.1. Physical and nutritional requirements of brewing yeast

Brewing yeast are mainly classified as the top-fermenting or ale yeast of *S. cerevisiae*, and the bottom-fermenting or lager yeast of *S. pastorianus*. These two yeast species can be differentiated in terms of their temperature of fermentation, sugar assimilation, genomic organization, evolutionary domestication, and phylogenesis. *S. cerevisiae* ferments between 18 and 24°C, with a maximum growth temperature of 37°C or higher; in contrast, *S. pastorianus* often ferments at lower temperatures, of between 8 and 14°C, although with a maximum growth temperature of 34°C. In general, brewing yeast requires high water activity, and in high sugar-containing wort they can overcome this stress condition through the overproduction of osmolytes, such as glycerol and trehalose, to protect their cell membranes. These osmolytes can replace water, to restore cell volume and osmotic pressure, and thus to allow regular yeast metabolism. For their pH requirements, yeast cells grow well between pH 4.5 and 6.5. During nutrient transport, yeast cells acidify their environment through a combination of proton secretion, direct secretion of organic acids, and CO_2 dissolution. Oxygen is required as a growth factor for the biosynthesis of their membrane fatty acids and sterols. As well as oxygen, yeast cells require the macronutrients (i.e., those needed at millimolar concentrations in the medium) of sulfur, free amino nitrogen, phosphorous, potassium, and magnesium. The micronutrients required by yeast cells (i.e., those needed at micromolar concentrations in the medium) are calcium, copper, iron, manganese, and zinc. *S. cerevisiae* requires low concentrations of growth factors, such as vitamins (e.g., biotin can be limiting), pyrimidines, purines, nucleotides, nucleosides, amino acids, sterols, and fatty acids. Generally, the malt wort provides growth factors, although in certain cases it can be necessary to supplement the wort with commercial yeast growth factors, as a mix of yeast extract, ammonium sulfate, and minerals (e.g., magnesium and zinc in particular). Barley malt wort is rich in maltose (50–60% total sugars), maltotriose (15–20%), glucose (10–15%), fructose and sucrose (1–2%, each), and dextrins (20–30%). These are derived from the hydrolysis of the starch by malt amylases during malting. However, *S. cerevisiae* cannot ferment dextrins, while *S. cerevisiae* and *S. pastorianus* show differences in terms of their melibiose fermentation, as only *S. pastorianus* can use this sugar (**Figure 1**).

The uptake of nutrients by yeast depends on the nutrient type, the yeast species, and the fermentation conditions. Generally, glucose is transported through the cell membrane into the cell by facilitated diffusion, and maltose by active transport. A high glucose concentration

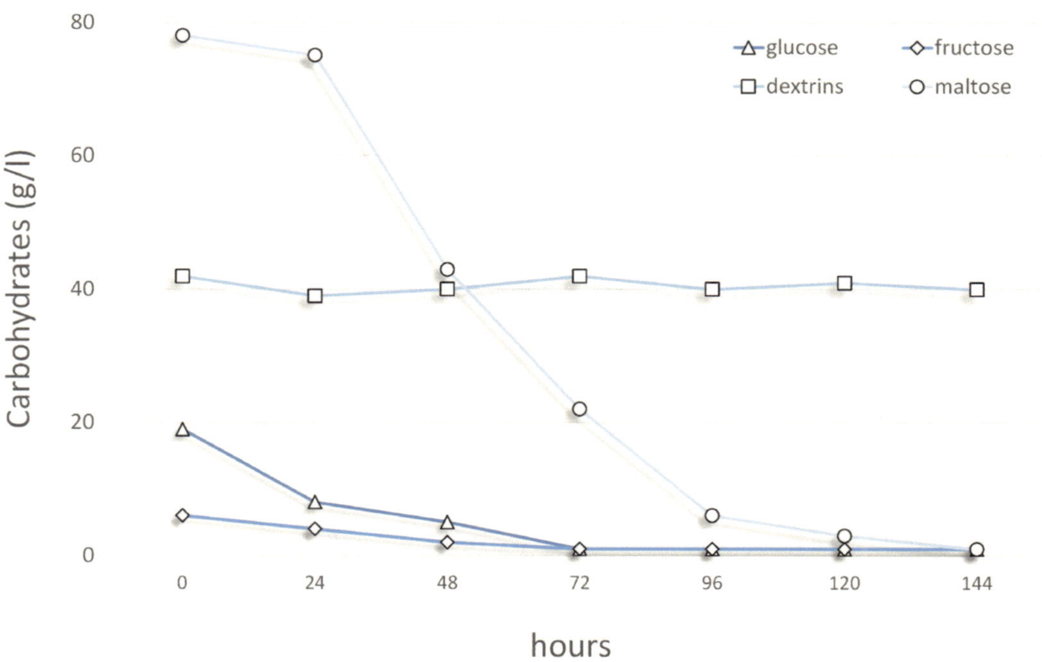

Figure 1. Carbohydrate uptake during wort fermentation (adapted from Ref. [2]).

in the wort can suppress the assimilation of maltose and other sugars (i.e., sugar catabolite repression). As a source of nitrogen, brewing yeast require assimilable organic (e.g., amino acids) and inorganic (e.g., ammonium salts) nitrogen for growth and fermentation. Again, high levels of ammonium ions in the wort can suppress the uptake of amino acids (i.e., nitrogen catabolite repression). Amino acid uptake occurs through two transport systems: general amino acid permease (GAP) and specific transporters for the different amino acids. The dissimilation of amino acids (i.e., decarboxylation, transamination, and fermentation) produces ammonium, glutamate, and higher alcohols (i.e., the fuel oils).

2.2. Genomic features of *S. cerevisiae* and *S. pastorianus* brewing strains

Over time, there was gradual domestication and selection of yeasts [3]. Moreover, there are two counteracting forces that act on yeast selection: yeast research needs to homogenize biological systems and to refer different yeast strains to species, while brewers need to differentiate and select yeast strains based on their fermentation characteristics. *S. cerevisiae* and *S. pastorianus* have followed different paths in their domestication that can now be read in their genome.

Ale strains of *S. cerevisiae* show wide heterogeneity, with differences in ploidy, genomic structure, and phenotypic behavior. *S. cerevisiae* has 5780 protein-encoding genes on 16 chromosomes (i.e., in haploids cells). However, the strains of *S. cerevisiae* that have been sequenced since 1996 represent laboratory yeast strains, and although these represent a well-characterized cell system, most of the known data are not adequate to understand and analyze the differences that characterize the diverse strains of ale yeast. *S. cerevisiae* ale strains can have different levels of ploidy, with some being aneuploid (i.e., with an abnormal chromosome number), and others being polyploid (i.e., with multiple complete genomes). Brewers have selected yeast strains over the centuries for the stability of their traits, and this has resulted in low spore viability and yeast that are deficient in sexual recombination.

Brewing strains belonging to *S. pastorianus* have been strictly linked to lager beer production for centuries, as seen by the fermenters and barrels of the central European brewers (e.g., Germany, Denmark, and Czech Republic). The origin of these strains has only recently been discovered through comparative genomic analysis, which revealed that *S. pastorianus* is a hybrid between *S. cerevisiae* cool-adapted strains and *Saccharomyces eubayanus* strains that have only been isolated from two possible wild reservoirs in Patagonia and Tibet. The production of lager beers started in central Europe around the end of the fourteenth century.

Recently, it was shown that on the basis of the sequence of different isolates of lager beer strains, *S. pastorianus* can be divided into two lineages [4]. One of these is associated with breweries in Denmark (i.e., Saaz-type, formerly *Saccharomyces carlsbergensis* triploid), which is characterized by a genome that is composed of most of the *S. eubayanus* genome (2n) and a partial *S. cerevisiae* genome (1n). The other lineage is from Germany (i.e., Frohberg-type, tetraploid) and is composed of equal genomes from *S. eubayanus* (2n) and *S. cerevisiae* (2n). Furthermore, the present-day *S. eubayanus* strains can also be hybrids themselves, with their genomes including portions of the genomes from *Saccharomyces uvarum*, the old *S. eubayanus*, and *S. cerevisiae*. In European wild environments, *S. eubayanus* has never been isolated, and this supports the hypothesis that *S. eubayanus* is a product of the lager-brewing environment, where these different species can be found growing together. Cold-tolerant yeast hybrids adapted well through serial passages during lager production, and these have become the dominant strains. Indeed, *S. pastorianus*, *Saccharomyces bayanus*, and *S. uvarum* have only been isolated from human-associated fermentation environments (**Figure 2**).

2.3. *Brettanomyces* species

Other species can contribute to wort fermentation and beer quality, including wild strains and species in open and uncontrolled fermentations. *Brettanomyces bruxellensis* is characteristic of the fermentation of Belgian lambic and geuze beers. Recently, genomic differences were reported for the *B. bruxellensis* contaminant yeast in wine (i.e., spoilage yeast) and the *B. bruxellensis* isolated from Belgian beers (i.e., brewing yeast). This study revealed that 20 genes in the spoilage strain genomes have been deleted in the brewing strains, many of which are involved in carbon and nitrogen metabolism. DNA fingerprinting has revealed that brewing strains have a unique profile, which means that they can be distinguish from spoilage strains. Thus, as for *Saccharomyces* strains, the selection undergone by *Brettanomyces* spp. also appears to have been influenced by adaptive modifications to brewing processes.

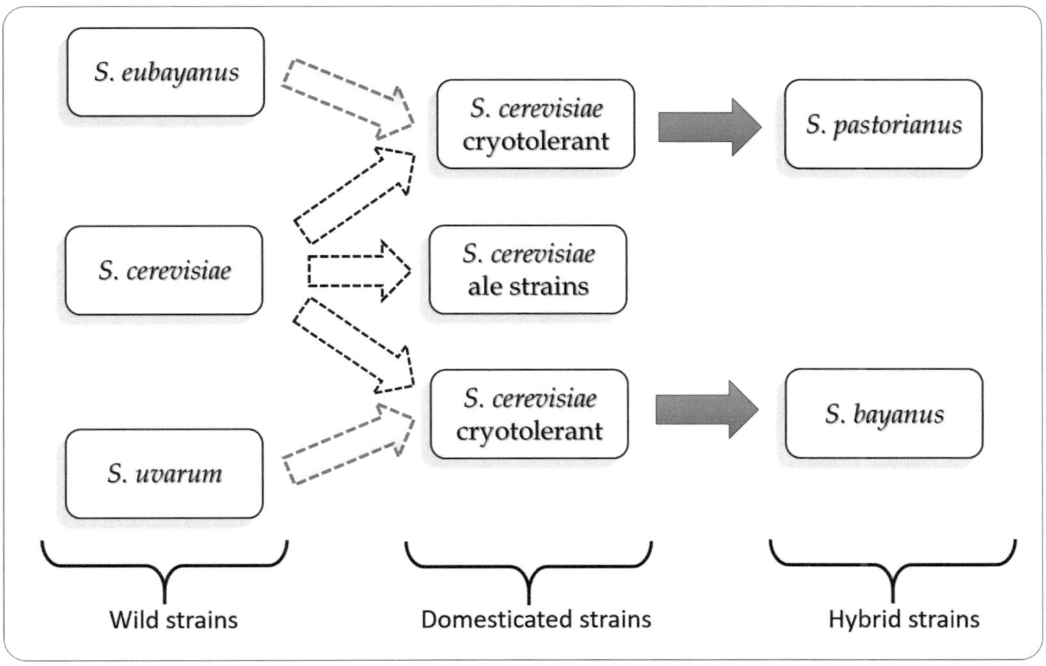

Figure 2. Source and selection of brewing yeast species (adapted from Ref. [5]).

3. Yeast management for aroma production

Generally, at the end of the boiling, the wort contains all of the nutrients that are required for yeast growth and fermentation. Thus, from a microbiological perspective, the main aspects to consider are not whether the wort is suitable for yeast growth, but rather what is the balance of the flavor compounds that will be produced by the yeast. Mathematical models have been developed to predict the final concentrations of some of these volatile compounds from the (known) quantities of their precursor(s) in the wort [6]. However, the application of these methods requires particularly deep knowledge of the wort composition, while brewers usually conduct very basic measurements in their evaluation of wort quality. For example, there are indications that small changes in the spectrum of the wort amino acid composition can result in dramatic changes in the final beer aroma. The most important metabolites synthesized by yeast and related to beer quality are sulfur compounds, organic and fatty acids, carbonyl compounds, higher alcohols, and esters.

3.1. Sulfur compounds

Sulfur compounds, such as hydrogen sulfide, methional, and dimethyl sulfide (DMS), are active flavor components of the beer that is generated during mashing and fermentation. During fermentation, through the metabolizing of amino acids and vitamins, and through the use of the inorganic components of the wort (e.g., sulfates), *S. cerevisiae* produces sulfur compounds, which include hydrogen sulfide. Sulfur compounds impart specific flavors to the beer, which have been defined as onion, rotted vegetables, or cabbage flavors, among others. While the over accumulation of these sulfur compounds is often undesirable, under specific circumstances, sufficient sulfite levels are necessary in the beer, to maintain flavor stability. Moreover, the sulfidic taste of dimethyl sulfide (below 100 μg/l) is an essential part of the flavor of lager beers and some ale beers. The major source of DMS in beer is the enzymatic conversion of dimethyl sulfoxide (DMSO) by yeast reductases. In particular, it has been shown that the disruption of the genes coding for methionine sulfoxide reductase abolished formation of DMS from DMSO in both *S. cerevisiae* and *S. pastorianus* [7, 8]. Methionine has a key role in the production of sulfur compounds by yeast. Indeed, when there are sufficient levels of methionine in the wort, this can cause inhibition of sulfite uptake, and the production of hydrogen sulfide [9].

3.2. Organic and fatty acids

The organic acids in beer are derived mainly from the yeast, as they are produced during the tricarboxylic acid, or Kreb's cycle (e.g., succinate and malate), from the catabolism of amino acids, and from redox reactions. Other organic acids, such as citrate and pyroglutamate, derive directly from the wort, and the yeast do not affect their concentrations in the beer. Overall, more than 100 organic acids have been identified in beers. These contribute to the reduction in pH during fermentation, and to the "sour" or "salty" taste of the beers. Fatty acids are of particular interest here, because of their involvement in the synthesis of esters. Yeast can incorporate saturated fatty acids and unsaturated fatty acids (UFAs) from the wort, or they can synthesize these from acetyl-CoA. However, the lack of sufficient oxygen in the later phases of fermentation makes the synthesis of UFAs impossible, and as a consequence, medium-chain fatty acids (MCFAs) are released into the medium [10]. These MCFAs are powerful detergents, and they can influence yeast vitality, beer taste, and foam stability. In particular, the typical flavor that is characteristic of MCFAs is defined as a "rancid goaty" flavor, and hence is often described as a "caprylic" flavor.

3.3. Carbonyl compounds

The presence of aldehydes and vicinal diketones is considered undesirable for beer quality. Acetaldehydes have unpleasant "grassy" flavors that are reminiscent of green apples and dry cider. In some circumstances, such as when there is excessive wort oxygenation and high pitching rates, aldehydes can accumulate in concentrations above the flavor threshold [11]. The vicinal diketones are important off-flavors of lager beers, which include diacetyl. During fermentation, yeast cells excrete an intermediate of valine biosynthesis, α-acetolactate, that is, spontaneously decarboxylated to diacetyl. Diacetyl has a strong aroma of toffee and butterscotch, and at concentrations above 0.05 ppm, it is considered as undesirable in lager beers. During conditioning, diacetyl is assimilated by yeast, and thus reduced to acetoin and 2-3 butandiol, which have much lower impact on beer quality. Traditionally, the rate-determining step of diacetyl accumulation in beer has been considered as the spontaneous decarboxylation of acetolactate, with yeast assimilation left with a marginal role. However, the physiological conditions of yeast are essential for diacetyl production and the time necessary for its reduction. High concentrations of valine and isoleucine in the wort inhibit vicinal diketone production by yeast. High assimilation rates have been observed at higher fermentation temperatures and when yeast is grown under aerobic de-repressed conditions. On the contrary, at higher pitching rates, the elevated concentrations of vicinal diketones produced by yeast require longer standing times [12].

3.4. Higher alcohols

The higher alcohols are also known as "fusel alcohols", and these are the most abundant organoleptic compounds in beer. Isoamyl alcohol, n-propanol, isobutanol, 2-phenyl-ethanol, and triptothol are important flavor and aroma components in terms of their concentrations. Below 300 mg/L, these compounds add complexity to the beer, by conferring refreshing, flower, and pleasant notes, and imparting a desirable warming character. On the contrary, above these concentrations, these compounds can have unpleasant heavy solvent-like odors. The formation of higher alcohols by brewing yeast involves different complex pathways, and a lot of progress has been made in the determination of the roles of the key genes involved in their biosynthesis [13]. The predominant idea for many years was that the higher alcohols are produced *via* the Ehrlich pathway. In this scheme, the yeast absorbs and deaminates the amino acids in the wort, with the resulting α-keto acids decarboxylated to aldehydes, and then reduced to higher alcohols. While this pathway can correctly explain the relationships between leucine and the corresponding isoamyl alcohol, it fails to explain why some fusel alcohols (e.g., n-propanol) do not correspond to any known amino acid, and that in wort containing low amino acid levels, there is no significant correlation between amino acids and higher alcohol composition of the beer. Indeed, in complex media such as the wort, most higher alcohols are formed following the glycolytic pathway. By fermenting the wort sugars, yeast not only produce ethanol, but also a number of long-chain alpha-acids that can subsequently be transformed into amino acids such as aspartate and glutamate. Finally, the choice of yeast strain can have great impact on higher alcohol production, and ale strains are considered to be higher producers than lager strains [13].

3.5. Esters

Esters are chemical compounds derived from a carboxylic acid and an alcohol, and they are of major industrial interest because they have very low thresholds and define the fruity aroma of the beer. Two main classes of esters are of particular interest for brewers: acetate esters and MCFA esters. Acetate esters have concentrations above threshold levels in most lager beers (e.g., isoamyl acetate is responsible for the banana-like aroma) and ale beers (e.g., ethyl acetate gives a solvent-like aroma). It is generally believed that the acetyl-CoA that is necessary for formation of acetate esters derives from oxidation of acetaldehyde. The acyl-CoAs required for the synthesis of MCFA esters originate from intermediates in the synthesis of fatty acids. Among the MCFA esters, ethyl hexanoate (i.e., an apple-like aroma) is an important flavoring compound, with levels above threshold in ale beers. The biosynthesis of esters requires acetyl-CoA or acyl-CoAs esterification with ethanol or higher alcohols by the specific alcohol acetyltransferase enzymes ATF1 and ATF2. Different studies have focused on the manipulation of fermentation conditions and the wort composition in ways that favor the availability of these factors and that lead to increased production of higher alcohols and esters.

4. Fermentation conditions

It is generally accepted that any condition that stimulates yeast growth will increase the production of higher alcohols and their acetate esters during fermentation. In this respect, increasing fermentation temperatures leads to an accumulation of acetate esters, with no significant differences in the levels of the MCFA esters. In particular, increasing the temperature from 10 to 12°C, increases ester production by up to 75% [14]. This phenomenon is dependent on increased alcohol acetyltransferase activity and stimulation of higher alcohol synthesis, which results from greater amino acid turnover. In addition, it has been suggested that higher fermentation temperatures increase the synthesis of a specific permease, Bap2p, that is, involved in import of the branched-chain amino acids valine, leucine, and isoleucine, which are known precursors of the higher alcohols [15].

Oxygen has an ambiguous role here. Indeed, oxygenation of the wort provides for better yeast growth, and consequently increased higher alcohol production. However, it is well-known that oxygenation leads to lower levels of esters in the beer. Oxygen acts in two different ways. First, availability of oxygen allows the biosynthesis of UFAs that is required to sustain yeast growth during fermentation. UFAs are esterified to glycerol to form membrane lipids, and thus, less acyl-CoA is available for synthesis of MCFA esters. Second, oxygen inhibits transcription of the alcohol-acetyltransferase-encoding gene *ATF1*, and consequently it reduces synthesis of acetate esters [13]. Thus, correct management of wort oxygenation at the time of pitching is essential to produce quality beers.

Another fermentation parameter that can affect ester synthesis is the hydrostatic pressure on the yeast cells. While in small craft breweries this is not a problem, the use of big cylindro-conical reactors in industrial beer production substantially increases the concentration of carbon dioxide dissolved in the beer. This has a double effect on ester synthesis. First, it inhibits yeast growth by lowering intracellular pH, and second, it directly reduces decarboxylation reactions, such as acetyl-CoA synthesis from pyruvate. As a consequence, large fermenters have been successfully used for the reduction of esters in beer, in particular during the fermentation of high gravity wort.

Stirring of the medium modulates the effects of the oxygen and carbon dioxide that are dissolved in the beer, to provide better oxygen distribution and decrease carbon dioxide supersaturation. Consequently, the synthesis of higher alcohols is stimulated, while that of MCFA esters is reduced, and the beer will have a less fruity aroma.

5. Wort composition

The amounts and types of sugars in the wort can influence the aromatic profile of the beer. Beers obtained from worts with higher percentages of glucose and fructose have higher ester levels than those obtained from maltose-rich worts. It has been suggested that glucose and fructose stimulate the glycolytic pathway and eventually lead to high levels of cytoplasmic acetyl-CoA, while maltose-rich wort only weakly induces acetyl-CoA formation [16]. Moreover, glucose induces *ATF1* and *ATF2* expression [13]. Similarly, high gravity brewing leads to the production of disproportionate amounts of ethyl acetate and isoamyl acetate, which give the beer over-fruity and solvent-like aromas, even after dilution to the standard ethanol content (i.e., 5% vol/vol).

It has been suggested that high nitrogen wort induces the transcription of *BAT1* and *ATF1*, thus, increasing both higher alcohols and acetate ester synthesis [16]. This effect is even more pronounced in the presence of elevated concentrations of valine, isoleucine, and leucine, which are known precursors of higher alcohol synthesis *via* the Ehrlich pathway.

The effects of free UFAs on beer aroma have been well documented. Similar to oxygen, low UFA levels in the wort increase ester synthesis, by recovering optimal yeast growth. However, at higher concentrations, UFAs relieve the need for the yeast to produce acetyl-CoA for lipid biosynthesis. This in turn induces lower MCFAs production, and lower production of their respective esters. Moreover, UFAs can directly repress *ATF1* transcription through the low oxygen response element [13]. Considering these effects, addition of UFAs has been proposed to modulate the final aroma of the beer obtained from a high gravity wort.

Finally, zinc stimulates the breakdown of α-keto acids to higher alcohols, thus, increasing their concentrations, and those of their corresponding esters.

5.1. Flocculation

At the end of the primary fermentation, yeast cells must be removed from the "green", or immature, beer. Flocculation is the process by which yeast cells aggregate and form "flocs" consisting of thousands of cells. *S. cerevisiae* ale strains rapidly separate from the beer by rising to the surface of the fermenter, probably adsorbed to carbon dioxide bubbles. On the contrary, *S. pastorianus* lager strains form flocs that sediment in the bottom of the fermenter. Yeast flocculation is an off-cost process of cell separation that does not require energy input. Given the importance of flocculation for the brewing industry, and to increase the efficiency of this process while avoiding premature flocculation, several mechanisms have been proposed to explain the physiological mechanisms involved in flocculation. In particular, the lectin-like mechanism is that generally accepted [17]. In this model, flocculent cells express lectin-like proteins that once activated by calcium ions, can recognize and bind to mannans in the cell wall of adjacent cells. This interaction is further stabilized by hydrogen bonds and hydrophobic interactions. In particular, positive correlation between cell hydrophobicity and flocculation has been demonstrated [18]. Transcriptional regulation and structural characteristics of the flocculation (*FLO*) genes that encode for lectin-like proteins are essential to produce yeast cells that can undergo flocculation. In particular, *FLO1* causes flocculation of a Flo1 phenotype, while a *FLO1* homolog, called Lg-*FLO1*, is responsible for the NewFlo phenotype observed in lager and ale brewing strains [19]. The NewFlo phenotype is modulated by sugars (i.e., maltose, glucose, mannose, and sucrose), extreme pH, oxygen availability, cations (e.g., Ba^{2+}, Sr^{2+}, and Pb^{2+}), and temperature. Sugars inhibit flocculation by competing with the binding sites of lectins, and by affecting the expression of the *FLO* genes. A pH outside the optimal range of pH 3–5 can affect yeast flocculation by interfering with the lectin conformation, by antagonizing the calcium activity, and by modifying the net electrical charge of the cell. Oxygenation of the medium represses expression of the anerobic cell wall mannoproteins and flocculation lectins [20].

Thus, on this basis, the nutritional and physiochemical conditions of sweet wort inhibit yeast flocculation. Indeed, the *FLO* genes are repressed and the yeast cell wall has a net negative charge that prevents cells from interacting with each other, and thus from aggregating. During fermentation, the decrease in pH, the prolonged anerobiosis, and the reduced availability of nutrients induce the NewFlo phenotype. In particular, lack of sugars (i.e., glucose, fructose, and maltose), or of nitrogen sources and lipids, can trigger flocculation of ale and lager yeast strains, respectively [21]. In addition, the increased ethanol concentration has a positive effect on yeast flocculation, by acting on the expression of the *FLO* genes and causing decreased cell–cell electrostatic repulsion. Finally, at the end of fermentation, triggering of the NewFlo phenotype strains occurs, because strains at the end of their exponential phase of growth are more flocculent [21].

As well as sugars, lipids, metal ions, and nitrogenous compounds, other minor wort components can influence the flocculation of yeast. In particular, complex polysaccharides that can induce premature flocculation have been identified. These polysaccharides can act as a bridge between cells by interacting with yeast lectin-like proteins. It has been suggested that the binding between these polysaccharides and yeast cells is mediated by cationic antimicrobial peptides [22]. These compounds are produced by barley to protect against microbial attack, or in response to fungal contamination during the malting process. Premature flocculation leads to incomplete attenuation of the wort, as aggregated cells cannot ferment the residual sugars of the wort.

Another major industrial problem is that flocculent strains can gradually lose their ability to flocculate, and thus eventually they will not form aggregates. It has been suggested that this phenomenon can be ascribed to genetic alterations during yeast multiplication and subsequent re-pitching [20]. In the case of insufficient flocculation, the yeast cells need to be removed at the end of fermentation by centrifugation or other separation techniques. However, yeast cells exposed to the stress associated with centrifugation have lower viability and vitality. This, in turn, can negatively affect beer quality and stability.

6. Novel starters for novel beer

The use of *Saccharomyces* strains in controlled fermentations over decades is essentially based on three main features: (a) high and efficient ethanol production; (b) use of fermentation as the preferential metabolic pathway, combined with the positive Crabtree effect (i.e., repression of respiration by glucose); and (c) higher tolerance to ethanol and other environmental stresses. The domination of the fermentation processes by the inoculated yeast (i.e., first fermentation and re-fermentation in craft beers) is fundamental to the aromatic profile of the final beer produced.

Over time, there has been gradual domestication and selection of yeasts [3]. Selected *Saccharomyces* strains are used for various purposes because of their plasticity for the assimilation of different substrates, which are usually not incorporated by *S. cerevisiae*. Indeed, choice of the yeast strain to use in the brewing process is also crucial to achieve a product that is valued by consumers and that has the required distinctive features and flavors. Also, yeast strains isolated from fermented foods other than beer can produce distinctive fermentative aroma profiles in beer [23, 24]. They can transform flavor precursors of the raw materials into more flavor-active compounds, which thus contribute to the final aroma of the beer [13, 25]. In this context, genetic strain-improvement strategies to enhance the fermentation efficiency and aromatic profiles of *S. cerevisiae* have been proposed [26].

6.1. Genetic improvement of brewing strains

High quality sequencing, *de-novo* assembly, and extensive phenotyping of 157 *S. cerevisiae* strains used for industrial production of beer and other fermented beverages (in their natural ploidy) have revealed that industrial yeast are genetically and phenotypically distinct from wild strains [27]. On this basis, there are many genetic approaches to the design of a superior yeast that can ferment and provide a particular style of beer. To obtain this, genetic modification approaches have mainly been applied to *S. cerevisiae*, such as rational metabolic engineering and inverse metabolic engineering. The aims here have been to increase the favored aromatic compounds or reduce undesired molecules. The nongenetic modification technique of hybridization can be used to increase the yeast fermentation fitness, ethanol tolerance, flocculation, and dextran degradation. Evolutionary engineering methods have mainly been applied to *S. pastorianus*, to improve fermentation capacity and flavor formation [28].

The positive role of *S. eubayanus* in the brewing process was only recently discovered through the use of hybridization tools [29]. Indeed, the natural interspecies *S. pastorianus* (i.e., *S. cerevisiae* × *S. eubayanus*) hybrid yeast is responsible for global lager beer production and is one of the most important industrial microorganisms [30]. Its success in the lager-brewing environment is due to a combination of traits that are not commonly found in *S. cerevisiae* yeast, as mainly the low temperature tolerance and maltotriose use. However, the hybrid origin of *S. pastorianus* and the presence of more genomes in one strain might affect the genetic improvement of these strains. To overcome this inconvenience, *de-novo S. cerevisiae* × *S. eubayanus* hybrids have been shown to outperform their parent strains in a number of respects, including, but not restricted to, fermentation rate, sugar use, stress tolerance, and aroma formation.

6.2. Non-*Saccharomyces* yeast

In the last few years, the selection of starter strains has also been carried out within non-*Saccharomyces* species that have been isolated and characterized as having distinctive aromatic and flavor components [11]. Indeed, the production of quality beer depends on the activity of the fermenting yeast that are selected not only for their good fermentation efficiency, but also for the characteristic aroma and flavors that they can give to the final product. In particular, the worldwide growth of craft beer has reinforced and encouraged the selection and use of different yeast genera, with pronounced impacts on aroma and flavor [31]. Indeed, non-*Saccharomyces* yeast represents a large source of biodiversity for the production of new beer styles, and they have the potential for wider application to other beverages and for other industrial applications. Within non-*Saccharomyces* yeast that can be used as pure starter cultures for the wort fermentation, different genera and species have been proposed, including *B. bruxellensis, Torulaspora delbrueckii, Candida shehatae, Candida tropicalis, Zygosaccharomyces rouxii, Lachancea thermotolerans, Saccharomycodes ludwigii,* and *Pichia kluyveri.*

7. Spontaneous fermentation

It is already known that some of the above mentioned yeast take part in the spontaneous fermentation processes of specialty beers. A typical example of this spontaneous process is the lambic beers. The fermentation of these beers is driven by brewery-resident microorganisms that are self-inoculated by exposing the wort in open tanks during the overnight cooling, before transferring it to wooden barrels for fermentation and aging. The fermentation of such Belgian lambic beers involves *Saccharomyce* spp. and *Brettanomyces* spp., with contributions from lactic acid bacteria and acetic acid bacteria [32]. Natural mixed fermentations are also used in the production of some German style "weiss" beers. Different yeast have been isolated during the maturation of acidic ale beers, including those that belong to the *Candida, Torulopsis, Pichia, Hansenula,* and *Criptococcus* genera, although their contributions to the aroma composition have not been well investigated. American coolship ale beer is another example of spontaneous fermentation, where the process can be divided into various steps: *Enterobacteriaceae* and oxidative yeast in the first phase, which are replaced by *Saccharomyces* spp. and *Lactobacillales*, and then *Dekkera bruxellensis*, which prevails through the final process [33]. Other nonconventional beers, such as Tchapalo, are brewed using *C. tropicalis* and *S. cerevisiae* cultures that have been selected for their fermenting of sorghum wort. In Africa, different types of beers are made, and one of these is known as "*Tchoukoutou*", which is a Beninese sorghum beer. The microbial ecology of the starter used to produce this beer demonstrated the presence of different species of non-*Saccharomyces* yeast, including *Pichia kudriavzevii, Candida ethanolica*, and *Debaryomyces hansenii* [34].

8. Nonconventional yeast

Recently, growing attention has been given to the possible contributions of nonconventional yeast to beer production. The success of craft beers has induced brewers to look for new alternatives for fermentation, such as nonconventional yeast, to impact on the aroma and flavor, and thus to generate differentiated products. The production and increase in the aroma compounds through biological methods exploits the metabolic pathways of the yeast for the promotion of the so-called bioflavor. This approach includes microbial bioconversion of the flavor precursors, use of strains that produce the required compounds, and genetic modification of the yeast [35]. In this regard, although still poorly investigated in the brewing sector, the use of nonconventional yeast might enhance the analytical and aromatic profiles of the final product and reduce the alcohol content [36].

Among the nonconventional yeast that can potentially be used in brewing, *T. delbrueckii* has received attention due to its fermenting of maltose, to produce ester compounds, and its biotransformation of the monoterpenoid flavor compounds of hops [25, 31, 36, 37]. More recently, it was studied the involvement of *T. delbrueckii* strains in the production of wheat style ("weiss") beers. In this investigation, it was demonstrated that *T. delbrueckii* can consume maltose more slowly than the *S. cerevisiae* commercial starter strain, while on the other hand, the nonconventional yeast give more intensity and complexity to the product [37]. In a recent study of the use of *T. delbrueckii* in wort fermentation Canonico and colleagues [36] looked at its influence on the analytical and aromatic profile of the beer, and the potential to produce low alcohol beer. These authors, after a preliminary screening among 28 strains, evaluated the use of a strain of *T. delbrueckii* in wort fermentation in pure and mixed cultures. The influence on the analytical and aromatic profile of beer, as well as the potential of producing a low alcohol beer using this strain of *T. delbrueckii* was evaluated. Results indicated that *T. delbrueckii* in mixed fermentation with a *S. cerevisiae* commercial starter can fully convert the fermentable sugars exhibiting distinctive analytical and aromatic profiles.

Moreover, Michel and colleagues [31] screened 10 *T. delbrueckii* strains on several brewing features, such us sugar use, hops, ethanol resistance, propagation, amino acid metabolism, and phenolic off-flavor-forming production, which revealed overall good fermentation of the wort and the production of a desirable fruity aroma. *L. thermotolerans* was also investigated for fermentation of the wort. Domizio and colleagues [38] tested three *L. thermotolerans* strains for important traits for beer production, including pitching rate, generational capacity, foam stability, hop tolerance, vicinal diketone production, oxygen requirement and flocculation, suggesting that *L. thermotolerans* may be a good choice for producing sour beers in a single fermentation step without the use of lactic acid bacteria.

8.1. Specialty beers

In response to increased consumer demand, the brewing industry has devoted much research effort to the development of new technologies and innovations for expansion of the assortment of specialty beers. Five types of specialty beers of particular interest have been described: low calorie beer, low alcohol or nonalcohol beer, novel-flavored beer, gluten-free beer, and functional beer [39]. Beers with a low calorie content have achieved great interest due to the problem of obesity, especially in Western populations, which accounts for a growing market segment.

This type of beer can be made by special mashing and collection of the wort with large amounts of fermentable sugars, or by inoculating microorganisms that can hydrolyze the more complex sugars, to reduce the concentrations of residual sugars in the final product. This is the case for *Brettanomyces/Dekkera*, the use of which leads to low calorie, but slightly more alcoholic, beers. Concerning this last aspect, new methods have been studied to produce beverages with lower alcohol contents, while providing superior sensory quality [40]. In this sense, non-*Saccharomyces* yeast represent a very attractive alternative, both for null processing of extra costs, and for the advantage of avoiding involuntary extraction of flavor compounds. *Wickerhamomyces subpelliculosus* (formerly *Pichia subpelliculosa*) and *Cyberlindnera saturnus* (formerly *Williopsis saturnus*), for example, have shown interesting results in the production of low alcohol beers with acceptable flavor profiles. *S. ludwigii* and the osmotic tolerant *Z. rouxii* are good examples of the novel use of non-*Saccharomyces* yeast to produce low alcohol or alcohol-free beers. *Pichia kluivery* was also recently proposed to produce positive flavor compounds from different hop varieties, thus indicating its potential application to the production of alcohol-free beer. Also, the use of *T. delbrueckii* strains that cannot degrade maltose, maltotriose, and other complex carbohydrates might be an interesting way to produce lower alcohol content in beers. The alcohol content here is usually close to 0.9% (v/v), with the advantage that many of these strains impart rich fruity flavor and aroma to the beer. The concept of low alcohol beer is necessarily linked to a functional product, and to a matrix with substantial amounts of fiber, vitamins, minerals, and polyphenols.

9. Functional beers

Functional beers are defined as beers with health benefits for those who consume them moderately. These are based on the use of nonconventional yeast that can produce or transform some beneficial compounds. This is the case for melatonin, which is a sleep-regulating hormone in mammals that has antioxidant properties and that can be produced in beer during alcoholic fermentation by the appropriate yeast [41]. Within functional beers, there are gluten-free beers for consumers with the condition known as coeliac disease, which is a gluten-sensitive and immune-mediated enteropathy. Recently, there has been increased demand and consumer interest to develop gluten-free beers from alternative cereals, such as sorghum and maize. In addition to those mentioned above, other yeast are promising candidates for the production of specialty beers. Indeed, yeast from the genera *Hanseniaspora, Pichia, Torulaspora, Wickerhamomyces,* and others, can offer diversified enzymatic and bioconversion, which are allowing brewers to work with new concepts that include bioflavoring and beers with reduced calorie and alcohol contents, or even functional beers.

10. Yeast handling in the brewery

One of the common and efficient cost-reduction measures in beer production is serial re-pitching of the yeast at the end of the fermentation. The type of fermenter and yeast used define the procedures needed for the recovery of the yeast biomass. Generally, by the time a crop has formed, the yeast should be removed as soon as possible, as it has no further positive role in the fermentation. The optimum time for yeast removal is usually decided by the brewers, by considering the different parameters, such as full attenuation of the wort, or the reduction of vicinal diketones to optimal levels.

Top-fermenting (ale) yeast are removed using specific skimming systems in the fermenters. Not all of the yeast heads that are cropped are retained for re-pitching. Only a fraction of the yeast head is used, as that composed of middle to young yeast cells and relatively free from trub. When using bottom-fermenting (lager) yeast, the yeast slurry on the base of the fermenting vessel is obtained by simply removing the overlying beer. As in the case of top-fermenting yeast, for subsequent use it is necessary to retain the yeast cells that are less enriched in trub and are of a middle age. This fraction can be identified in the middle of the sediment, as the lowest layers are enriched in trub. Furthermore, some authors have suggested that the more flocculent portion of the cells settle in the middle, while the less flocculent cells are in the top portion of the cone. In the majority of breweries, pitching yeast is usually converted to a liquid slurry by adding water or by leaving enough entrained beer to facilitate yeast transport to the storage vessels via pumping. Alternatively, yeast cakes can be obtained by recovering the yeast from the entrained beer through filtering.

10.1. Yeast storage

During storage, yeast quality decreases as a function of storage conditions and procedures. This can lead to aberrant fermentation, which can be seen as slow attenuation rates, poor flocculation performance, and undesired flavor development. In particular, it is necessary to avoid yeast contamination with bacteria, wild yeast, or other starter yeast (cross-contamination). A common procedure to reduce the bacterial load is to treat the yeast slurry with a chemical disinfectant at low pH (i.e., acid washing). Usually citric and phosphoric acids are used at low temperatures (2–4°C) with continuous gentle stirring [42]. The correct procedures during acid washing allow the removal of bacteria without affecting the yeast performances in the subsequent fermentation.

Another aspect of primary importance is the avoidance of excessive stress to the yeast cells during cropping and storage, to minimize any changes in their physiological conditions. In particular, the intracellular concentrations of storage carbohydrates (i.e., glycogen and trehalose) and sterols and other lipids are of primary importance for the duration of storage [43]. Indeed, the storage phase is a period of starvation, and the yeast need to rely on nutritional reserves that were accumulated during fermentation. Glycogen is synthesized during mid-fermentation, and its dissimulation is directly correlated with the storage temperatures. Furthermore, supplementation of cropped yeast with linoleic acid before pitching has been suggested as a convenient way to improve the yeast physiology without affecting the yeast growth, fermentation rate, and production of volatile compounds during the subsequent beer production [44].

10.2. Continuous re-pitching

The influence of serial cropping and re-pitching on the consistency of fermentation performance and beer composition poses technological questions about the number of generations that can be allowed to elapse before introducing new yeast. This decision is usually made by the individual brewers, as there are no pre-determined rules. In breweries with high hygiene standards, serial re-pitching can continue for 15–20 generations, while in microbreweries, even 5–10 generations is considered excessive. Continued serial re-pitching of yeast can be associated with gradual deterioration in the yeast conditions, which can result in decline in fermentation performance. Indeed, the aging process in yeast is associated with gradual disruption of many of their metabolic processes. On the other hand, there is an economic cost to propagation, and if a re-pitched yeast is performing satisfactorily, there is less need to introduce a newly propagated yeast. Indeed, the first generation fermentation using new yeast lines is atypical, as the yeast cells are less adapted to the wort and fermentation conditions, particularly for high gravity brewing. To assist brewers in the decision of when to introduce a new yeast line, several methodologies have been developed. The most widely used assay to evaluate yeast viability involves microscopic observation of the yeast cells stained with methylene blue. While this method is economic and simple to perform, it is also subject to operator error and known to overestimate viability [45]. Recently, flow-cytometric methods have been developed to assist brewers in the evaluation of their yeast viability and vitality. In particular, analysis of the yeast cells stained with the fluorescent dye oxonol allows automatic detection of yeast viability without interference from the wort trub [46]. Moreover, flow cytometry assays are not limited to viability tests, but can also be used for vitality tests that are related to the yeast fermentation performance, which can be implemented using specific fluorophores.

Author details

Marilena Budroni[1]*, Giacomo Zara[1], Maurizio Ciani[2] and Francesca Comitini[2]

*Address all correspondence to: mbudroni@uniss.it

1 Department of Agricultural Science, University of Sassari, Sassari, Italy

2 Department of Life and Environmental Sciences, Polytechnic University of Marche, Ancona, Italy

References

[1] Lodolo EJ, Kock JL, Axcell BC, Brooks M. The yeast Saccharomyces cerevisiae—the main character in beer. FEMS Yeast Research. 2008;**8**(7):1018-1036

[2] Stewart GG. Saccharomyces species in the production of Beer. Beverages 2016;**2**(34):1-18

[3] Sicard D, Legras JL. Bread, beer and wine: Yeast domestication in the Saccharomyces sensu stricto. Comptes Rendus Biologies. 2011;**334**(3):229-236

[4] Peris D, Sylvester K, Libkind D, Gonçalves P, Sampaio JP, Alexander WG, Hittinger C.T. Population structure and reticulate evolution of Saccharomyces eubayanus and its lager-brewing. Molecular Ecology. 2014;**23**(8):2031-2045

[5] Bird RC, Smith B.F. The impact of yeast genomics on brewing. Journal of Nutrition and Food Sciences. 2016;**6**(1):1-6

[6] Gee DA, Ramirez WF. A flavour model for beer fermentation. Journal of the Institute of Brewing. 1994;**100**:321-329

[7] Hansen J. Inactivation of MXR1 abolishes formation of dimethyl sulfide from dimethyl sulfoxide in Saccharomyces cerevisiae. Applied and Environmental Microbiology. 1999;**65**(9):3915-3919

[8] Hansen J, Bruun SV, Bech LM, Gjermansen C. The level of MXR1 gene expression in brewing yeast during beer fermentation is a major determinant for the concentration of dimethyl sulfide in beer. FEMS Yeast Research. 2002;**2**(2):137-142

[9] Landaud S, Helinck S, Bonnarme P. Formation of volatile sulfur compounds and metabolism of methionine and other sulfur compounds in fermented food. Applied Microbiology and Biotechnology. 2008;**77**(6):1191-1205

[10] Bardi L, Cocito C, Marzona M. Saccharomyces cerevisiae cell fatty acid composition and release during fermentation without aeration and in absence of exogenous lipids. International Journal of Food Microbiology. 1999;**47**(1-2):133-140

[11] Vanderhaegen B, Neven H, Coghe S, Verstrepen K J, Derdelinckx G, Verachtert H. Bioflavoring and beer refermentation. Applied Microbiology and Biotechnology. 2003;**62**(2):140-150

[12] Krogerus K, Gibson BR. 125th anniversary review: Diacetyl and its control during brewery fermentation. Journal of the Institute of Brewing. 2013;**119**(3):210-225

[13] Pires EJ, Teixeira JA, Brányik T, Vicent AA. Yeast: The soul of beer's aroma—a review of flavour-active esters and higher alcohols produced by the brewing yeast. Applied Microbiology and Biotechnology. 2014;**98**:1937-1949

[14] Verstrepen KJ, Derdelinckx G, Dufour J, et al. Flavor-active esters: Adding fruitiness to beer. Journal of Bioscience and Bioengineering. 2003;**96**(2):110-118

[15] Didion T, Grauslund M, Kielland-Brandt MC, Andersen HA. Amino acids induce expression of BAP2, a branched-chain amino acid permease gene in Saccharomyces cerevisiae. Journal of Bacteriology. 1996;**178**:2025-2029

[16] Lei H, Zhao H, Yu Z, Zhao M. Effects of wort gravity and nitrogen level on fermentation performance of brewer's yeast and the formation of flavor volatiles. Applied Biochemistry and Biotechnology. 2012;**166**(6):1562-1574

[17] Miki BL, Poon NH, James AP, Seligy VL. Possible mechanism for flocculation interactions governed by gene FLO1 in Saccharomyces cerevisiae. Journal of Bacteriology. 1982;**150**(2):878-889

[18] Jin Y, Ritcey LL, Speers RA. Effect of cell surface hydrophobicity, charge, and zymolectin density on the flocculation of Saccharomyces cerevisiae. Journal of the American Society of Brewing Chemists. 2001;**59**:1-9

[19] Stratford M, Assinder S. Yeast flocculation: Flo1 and newFlo phenotypes and receptor structure. Yeast. 1991;**7**:559-574

[20] Soares EV. Flocculation in Saccharomyces cerevisiae: A review. Journal of Applied Microbiology. 2011;**110**(1):1-18

[21] Verstrepen KJ, Derdelinckx G, Verachtert H, Delvaux FR. Yeast flocculation: What brewers should know. Applied Microbiology and Biotechnology. 2003;**61**:197-205

[22] Panteloglou AG, Smart KA, Cook DJ. Malt-induced premature yeast flocculation: Current perspectives. Journal of Industrial Microbiology and Biotechnology. 2012;**39**(6):813-822

[23] Marongiu A, Zara G, Legras JL, Del Caro A, Mascia I, Fadda C, Budroni M. Novel starters for old processes: Use of Saccharomyces cerevisiae strains isolated from artisanal sourdough for craft beer production at a brewery scale. Journal of Industrial Microbiology and Biotechnology. 2015;**42**(1):85-92

[24] Mascia I, Fadda C, Dostálek P, Karabín M, Zara G, Budroni M, Del Caro A. Is it possible to create an innovative craft durum wheat beer with sourdough yeasts? A case study. Journal of Institute of Brewing. 2015;**121**(2):283-286

[25] King A, Dickinson RJ. Biotransformation of monoterpene alcohols by Saccharomyces cerevisiae, Torulaspora delbrueckii and Kluyveromyces lactis. Yeast. 2000;**16**(6):499-506

[26] Steensels J, Verstrepen KJ. Taming wild yeast: Potential of conventional and nonconventional yeasts in industrial fermentations. Annual Review of Microbiology. 2014;**68**(1): 61-80

[27] Gallone B, Steensels J, Prahl T, Soriaga L, Saels V, Herrera-Malaver B, Teiling C. Domestication and divergence of Saccharomyces cerevisiae beer yeasts. Cell. 2016;**166**(6):1397-1410

[28] Procopio S, Brunner M, Becker T. Differential transcribed yeast genes involved in flavour formation and its associated amino acid metabolism during brewery fermentation. European Food Research and Technology. 2014;**239**:421-439

[29] Hebly M, Brickwedde A, Bolat I, Driessen MR, de Hulster EA, van den Broek M, Daran-Lapujade P. S. cerevisiae× S. eubayanus interspecific hybrid, the best of both worlds and beyond. FEMS Yeast Research. 2015;**15**(3):1-12

[30] Krogerus K, Magalhães F, Vidgren V, Gibson B. Novel brewing yeast hybrids: Creation and application. Applied Microbiology and Biotechnology. 2017;**101**(1):65-78

[31] Michel M, Kopecká J, Meier-Dörnberg T, Zarnkow M, Jacob F, Hutzler M. Screening for new brewing yeasts in the non-Saccharomyces sector with Torulaspora delbrueckii as model. Yeast. 2016;**33**(4):129-144

[32] Spitaels J, Freek L. The microbial diversity of traditional spontaneously fermented lambic beer. PloS One. 2014;**9**(4):e95384

[33] Bokulich NA, Charles WB, Mills DA. Brewhouse-resident microbiota are responsible for multi-stage fermentation of American coolship ale. PLoS One. 2012;**7**(4):e35507

[34] Basso RF, Alcarde AR, Barbosa C. Could non-Saccharomyces yeasts contribute on innovative brewing fermentations? Food Research International. 2016;**86**:112-120

[35] Mertens S, Steensels J, Saels V, De Rouck G, Aerts G, Verstrepen KJ. A large set of newly created interspecific Saccharomyces hybrids increases aromatic diversity in lager beers. Applied and Environmental Microbiology. 2015;**81**:8202-8214

[36] Canonico L, Agarbati A, Comitini F, Ciani M. Torulaspora delbrueckii in the brewing process: A new approach to enhance bioflavour and to reduce ethanol content. Food Microbiology. 2016;**56**:45-51

[37] Tataridis P, Kanelis A, Logotetis S, Nerancis E. Use of non-Saccharomyces Torulaspora delbrueckii yeast strains in winemaking and brewing. Journal of Natural Science. 2013;**124**:415-426

[38] Domizio P, House JF, Joseph CML, Bisson LF, Bamforth CW. Lachancea thermotolerans as an alternative yeast for the production of beer. Journal of Institute of Brewing. 2016;**122**:599-604

[39] Yeo HQ, Liu SQ. An overview of selected specialty beers: Developments, challenges and prospects. International Journal of Food Science and Technology. 2014;**49**(7):1607-1618

[40] Sohrabvandi S, Seyyed Hadi R, Seyyed Mohammad M, Amir M. Characteristics of different brewer's yeast strains used for nonalcoholic beverage fermentation in media containing different fermentable sugars. International Journal of Biotechnology. 2010;**8**:178-185

[41] Maldonado MD, Moreno H, Calvo JR. Melatonin present in beer contributes to increase the levels of melatonin and antioxidant capacity of the human serum. Clinical Nutrition. 2009;**28**(2):188-191

[42] Simpson WJ, Hammond JRM. The response of brewing yeasts to acid washing. Journal of Institute of Brewing. 2013;**95**(5):27-32

[43] Quain DE, Tubb RS. The important of glycogen in brewing yeast. Technical Quarterly - Master Brewers Association of the Americas. 1982;**19**:29-33

[44] Moonjai N, Verstrepen KJ, Shen HY, Derdelinckx G, Verachtert H, Delvaux FR. Linoleic acid supplementation of a cropped brewing lager strain: Effects on subsequent fermentation performance with serial repitching. Journal of Institute of Brewing. 2003; **109**(3):262-272

[45] O'Connor-Cox E, Mochaba FM, Lodolo EJ, Majara M, Axcell B. Methylene blue staining: Use at your own risk. Technical Quarterly—Master Brewers Association of the Americas. 1997;**34**:306-312

[46] Boyd AR, Gunasekera TS, Attfield PV, Simic K, Vincent SF, Veal DA. A flow-cytometric method for determination of yeast viability and cell number in a brewery. FEMS Yeast Research. 2003;**3**(1):11-16

Chapter 5

Use of Non-*Saccharomyces* Yeasts in Bottle Fermentation of Aged Beers

María Jesús Callejo, Carmen González and
Antonio Morata

Additional information is available at the end of the chapter

http://dx.doi.org/10.5772/intechopen.68793

Abstract

Bottle fermented and brewed beers are reaching more recognition in present days due to their high sensory complexity. These beers normally are produced by an initial tank fermentation to metabolize the sugars obtaining the typical alcoholic degree, and later the foam and CO_2 pressure is produced by subsequent bottle fermentation. The sensory profile is improved by the formation of some fermentative volatiles, but also by the ageing on lees, because beers are brewed during several months with the yeast cells that performed the fermentation. The use of non-*Saccharomyces* yeast is a trending topic in many fermentative food industries (wines, beer, bread, etc.). They open new possibilities to modulate flavor and other sensory properties during fermentation and biological ageing. This chapter review the effect of some non-*Saccharomyces* yeasts such as *Schizosaccharomyces pombe, Torulaspora delbrueckii, Lachancea thermotolerans, Saccharomycodes ludwigii,* and *Brettanomyces bruxellensis* in the bottle fermentation and brewing of beers analyzing their metabolic specificities and sensory contribution on beer taste.

Keywords: non-*Saccharomyces* yeasts, beer, bottle fermentation, *Schizosaccharomyces pombe, Torulaspora delbrueckii, Lachancea thermotolerans, Saccharomycodes ludwigii, Brettanomyces bruxellensis*

1. Introduction

The beer-like beverages were already produced in the regions of Mesopotamia and Ancient Egypt since 5500 BC [1]. Home-made or small-scale trade brewing supplied an essential part of diet to a primarily agrarian population. The historical development of brewing and the brewing industry is, however, linked with northern Europe where cold conditions inhibited the development of viticulture.

Although the main malted cereal for brewing is barley, other cereals such as oats, maize, rice, rye, sorghum, and wheat are malted for specific purposes, mainly for the production of peculiar beers. From the tenth century, adding of hops became usual from Germany across Europe to replace, or at least supplement, the variety of plants, herbs, and spices popular at that time. Not only the pleasing flavor and aroma of hop but perhaps, more importantly, their action in protecting the beer from being spoiled by the then unknown microbes, eventually led to their wide-scale adoption.

According with Anderson [2], in the medieval and early modern period the scale of brewing ranged from a few hectoliters annually in the average home to hundreds, or occasionally, thousands of hectoliters in the largest monasteries and country houses. Domestic brewing still accounted for well over half of the beer produced at the end of the seventeenth century. On the eighteenth and nineteenth centuries, the population of Europe's cities growth was accompanied by an increase in beer consumption. The leading European beer-drinking countries of nineteenth century were the United Kingdom, Germany, and Belgium. Large-scale production began on the nineteenth century with improvements of technology and scientific research on microbiology. First, little meaningful scientific research in brewing was carried out by Louis Pasteur whose investigations on wine and beer fermentations in the 1860s and 1870s showed the importance of eliminating deleterious bacteria [3]. Emil Christian Hansen became the first to isolate a pure yeast culture at Carlsberg in 1883.

The twentieth century saw the expansion in the beer industry. The industrial cooling allowed the introduction into the market of lager beers. Currently, there is a resurgence in craft or microbrewers [4], the market of craft beers in the United States grew 12.8% in 2015, and a significant strengthening of the sector has also been detected in several European countries, where growth from 2009 to 2014 outstripped 90–145% in Switzerland, the United Kingdom, France, Italy, and Slovenia; 333–400% in Slovakia, Czech Republic, Sweden, and Norway; and more than 1000% in Spain [5].

Brewing can be defined as the making of beer or related beverages by infusion, boiling, and fermentation. The various processes, inputs, and products of a brew house are shown in **Figure 1**. First part of the brewing process consist in converting the raw materials—water, malt, adjuncts, and hops—into a fermentable wort. Malt enzymes assist in the conversion of starch to fermentable sugars by mashing. For that reason, prior to the brewing process the malting process activates the natural enzyme systems of barley by controlled steeping, germination, and kilning. Mashing was simply the process of mixing warm water with ground malt, and cereal adjuncts if used, and after a period of standing, as much of the liquid as possible was recovered. The mash is then transferred to a mash filter (lauter tun) to produce bright wort and to collect the maximum amount of sugars (extract) from the residual solid materials ("spent" grain). Wort boiling satisfies a number of important objectives such as sterilization of the wort, extraction of the bittering compounds from hops, coagulation of excess proteins and tannins to form solid particles (hot trub) that can be removed later in the whirlpool, color and flavor formation, removal of undesirable volatiles such as dimethyl

http://dx.doi.org/10.5772/intechopen.68793

sulfide (DMS), by evaporation, and the concentration of the sugars by evaporation of water. Hot trub does need to be removed if the beer stability is not to suffer. The wort is then cooled from almost boiling point to fermentation temperature through a heat exchanger using water as the main cooling medium. The temperature for fermentation (**Figure 2**) is different for ale

Figure 1. Malting and brewing process. First steps mashing and wort preparation.

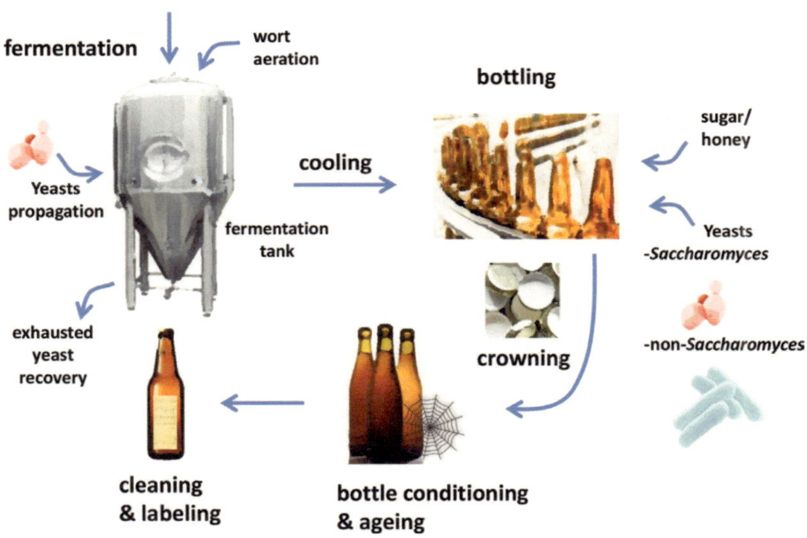

Figure 2. Brewing process: fermentation and bottle fermentation.

or lager fermentation (typically 8–14°C for lager and 15–20°C for ale). Yeast is pitched into the cooled wort in the fermentation vessel. Once the fermentation of the wort is completed, it is important to remove the bulk of the excess yeast before maturation, generally by removing the beer from the settled yeast. Then green beer must be conditioned to produce a stable, quality product suitable for filtration, and packaging. This process is called aging (lagering in lager beers). The objectives of beer aging are chill haze formation, clarification, carbonation (to a limited extent), and flavor maturation (again to a limited extent).

Fermentations performed in closed cylindro-conicals vessels may be of either ale or lager type. At the end of primary fermentation, the bulk of the yeast is collected by the application of rapid chilling and we have a green beer. Maturation of green beer is needed to obtain flavor adjustment (diacetyl, SO_2, and DMS), yeast sedimentation, carbonation, and colloidal stability.

Typically, ale fermentations are fully attenuated and subjected to a short low-temperature conditioning in a separate tank to adjust carbonation, precipitate chill haze, and allow some loss of undesirable flavor volatiles trough gas purging. As lager beers ferment at much lower temperatures than ale yeasts (and therefore much slower), they also require an extended period of cellaring for maturation and development of the beer flavor. Conditioning is carried out at low temperature for no longer than a few days. As with chilled and filtered ales, this part of the process serves simply to adjust carbonation, develop chill haze protection, and clarify the beer. After conditioning the beers are filtered and packaged.

Other recovering old production methods for innovation are the production of cask- and bottle-conditioned beers. With bottle-conditioned beers a base beer is used. This base beer are as nearly as possible completely attenuated. This allows accurate control of addition of priming sugars. After bottling, beers are held in the brewery for a period from a few weeks to a few months depending on the temperature. Products of yeast metabolism are excreted into the beer providing distinctive flavor [4]. It is possible to apply this technique by adding pure cultures of different non-*Saccharomyces* yeast strains to increase the complexity of the final product.

2. Fermentation

From ancient times the elaboration of beer was carried out by spontaneous fermentations with exposure to ambient to air of the cereal slurry to induce "contamination." Bit-by-bit species of the genus *Saccharomyces* were intuitively selected from different raw materials. Nowadays, in most beers, the microorganisms that carry out fermentation belong to different *Saccharomyces* species and strains [6]. For 99% of the worldwide beers, *Saccharomyces* is the sole microbial inoculum.

According to Ref. [4], the use of *Saccharomyces* strains in controlled fermentations over decades is essentially based on three main features: (a) efficient production of high ethanol amounts; (b) the use of fermentation as the preferential metabolic pathway, combined to the positive

Crabtree effect (repression of respiration by glucose); and (c) higher tolerance to ethanol and other environmental stresses [7].

In brewing most of the strains of *Saccharomyces* are classified into the categories such as ale and lager yeasts. Although there are many different styles of beer, the main brewing classification criterion particularly relies on the selection of the yeast strain and type of fermentation. Ale yeasts or top-fermenting yeasts, which are *Saccharomyces cerevisiae* strains, rise up to the surface of the vessel with the escaping carbon dioxide gas bubbles and become entangled in the fermentation head, facilitating their collection by skimming. Fermentation temperature ranges between 15 and 20°C. Lager yeasts or bottom-fermenting yeast, do not rise and become entrapped in the foam but settle out at the end of the fermentation. The nomenclature of lager yeast has evolved as research has been developed. Successively, they have received different names such as *Saccharomyces carlsbergensis, Saccharomyces uvarum*, and *S. cereviseae* lager type. Now they are termed *S. pastorianus* [6]. Lager worts often ferment at lower temperatures (8–14°C) than ale yeasts and are therefore much slower.

The occurrence of other species different to *Saccharomyces* species and strains is commonly reported in some peculiar beer styles produced by spontaneous fermentations, as the Belgian acid beers (Lambic, Gueuze, and rodenbach), and the American Coolship Ales, an American descendant of the Belgian lambic style. In these processes, wort is spontaneously fermented by microbes present in the air and surfaces of the brewery. These microorganisms are introduced by exposing the wort in shallow tanks during the overnight cooling, before transferring it to wooden barrels for fermentation, and aging.

The non-*Saccharomyces* species can be used in the production of low-alcohol beer (0.5–1.2%, v/v) and alcohol-free beer (<0.5%, v/v) [8]. The suitability of yeast strains of *Saccharomycodes ludwigii* and *Zygosaccharomyces rouxii* for low-alcohol beer production has been analyzed by De Francesco et al. [9]. Most of the *Z. rouxii* strains were found unsuitable because of the production of high concentration of ethanol. The most successful genus used for the industrial production of alcohol-free beer is *S. ludwigii* due to its disability to ferment maltose and maltotriose, the prevailing fermentable sugars of all malt [10, 11].

Other application of non-*Saccharomyces* recently emerged is its use in controlled fermentations. This practice has been gaining popularity among brewers in order to obtain distinctive products, with distinctive aromatic and flavor components [12].

Although there are different options of innovation in the brewing process (using special malts or adjuncts, hop varieties, water quality, etc.), new and novel brewing yeasts strains can be discovered to enhance the aroma and flavor characteristics of beer and provide opportunities for developing new beers. The production of most aroma-active compounds is strictly dependent on the yeast strain chosen for the fermentation [13, 14]. This makes the selection of suitable strains the most important task to make good beer and opens new innovation opportunities to gain market especially in the market of craft-beer, to improve aroma profile after inoculation of non-*Saccharomyces* yeasts during bottle conditioning and ageing.

3. Use of non-*Saccharomyces* in beer fermentation and ageing

The use of non-*Saccharomyces* yeasts allows new possibilities in the improvement and innovation in sensory profile of beers but also technological advantages and new ways of ageing. Several articles review the importance of non-*Saccharomyces* yeasts in beer [4, 15] and other fermented beverages [16, 17].

The typical yeast species used in beer fermentation, and in most of food fermentations, is *S. cerevisiae* which is able to ferment monosaccharides like glucose, fructose, galactose, and mannose, and some disaccharides such as maltose and sucrose and also is able to use the trisaccharide rafinosse [18]. Some strains can metabolize more than 300 g/L of sugars by fermentation reaching 18% (v/v) of ethanol in some alcoholic beverages like wines. As nitrogen source one can use urea, ammonia, and several amino acids. Also phosphate, biotine, and other cofactors and micronutrients are also needed. *S. cerevisiae* has globous shape with multipolar budding and is teleomorphic yeast with tetrahedral sporulation that can be promoted in acetate agar.

Beer yeast can be classified in top and bottom-fermenting yeasts even when both belong to *S. cerevisiae* species. Ale strains ferment better at 15–20°C, a relatively warm temperature, and during fermentation they form a thick film on the surface formed by yeast cells. They are considered top-fermenting yeasts which forms high amounts of esters that produce the distinctive sensory profile of ale beers such as ales, porters, stouts, Altbier, Kölsch, and wheat beers. Lager strains works better at 8–14°C growing slowly and settling at the bottom of the tank, they are known as bottom-fermenting yeasts. Sensory profile depends a lot in the strain used, some famous lagers are Pilsners, Dortmunders, Märzen, Bocks, and American malt liquors.

The use of non-*Saccharomyces* yeasts open new possibilities compared with the traditional *S. cerevisiae* strains. In fact some traditional beers such as Lambic, are produced by spontaneous fermentations with the development of non-*Saccharomyces* yeasts specifically *Brettanomyces bruxellensis*. Their distinctive sensory profile is due to its metabolomic impact by the production and release of several esters with fruity smells such as ethyl acetate, ethyl caprate, ethyl caprylate, and ethyl lactate [4, 7, 15]. Moreover, the volatile acidity produces a sour taste typical in Lambic beer [19].

The main fermentative properties of interesting non-*Saccharomyces* yeasts reported as useful for beer and other fermented beverages are described in **Table 1**. *Torulaspora delbrueckii, Lachancea thermotolerans*, and *Schizosaccharomyces pombe* are currently produced at industrial level as dry yeast or liquid refrigerated starters by international biotechnological companies such as Chr. Hansen, Lallemand, Laffort, and Erbslöh [17].

One of the most studied non-*Saccharomyces* in modern food fermentations is *T. delbrueckii*, formerly known as *Saccharomyces delbrueckii* or *Saccharomyces rosei*. It is now being used to improve fermentation or technological properties in beer [4], wine [20], bread [21], and other food products, and it is possible to find commercial cultures as dry yeasts [17]. The morphology is ellipsoidal similar to *S. cerevisiae* (**Figure 3A**). *T. delbrueckii* is teleomorph yeast corresponding to the anamorph species *Candida colliculosa*. Typical sporulation form one or two

http://dx.doi.org/10.5772/intechopen.68793

Yeast species	Fermentative power (% v/v ethanol)	Able to ferment	Volatile acidity (g/L)	Volatile compounds	Effect on acidity	Ageing on lees	References
S. cerevisiae	12–18	Glucose Fructose Galactose (v) Sucrose Maltose	<0.5	Higher alcohols Esters	Neutral	Depending on strains	[16]
Torulaspora delbrueckii	<9	Glucose Fructose Galactose (v) Sucrose (v) Maltose (v)	<0.5	Ethyl lactate 2-phenylethyl acetate 3-ethoxy propanol	Neutral	Moderate-high	[4, 23, 24]
Lachancea thermotolerans	<9	Glucose Fructose Maltose (v) Galactose (v)	<0.5	2-phenylethyl acetate Ethyl lactate	Acidity enhancement lactic acid production	Depending on strains	[25]
Schizosaccharomyces pombe	12–14	Glucose Fructose Sucrose Maltose	0.8–1.4	Higher alcohols Esters	Maloalcoholic deacidification	Osmophilic high release of cell polysaccharydes	[16, 26, 27]
Saccharomycodes ludwigii	12–14	Glucose Fructose Sucrose	<0.5	Diacetyl, acetoin	Neutral	Osmophilic high release of cell polysaccharydes	[9, 28]
Dekkera (*Brettanomyces*) *bruxellensis*	<2	Glucose Fructose Sucrose Maltose Cellobiose	>1	Ethylphenols Isovaleric acid Isobutyric acid Pyrazines	Acidity enhancement	Release of cell polysaccharydes	[4, 15, 19, 29–31]

Table 1. Technological performance and features of non-*Saccharomyces* yeasts with potential advantages in brewing technology.

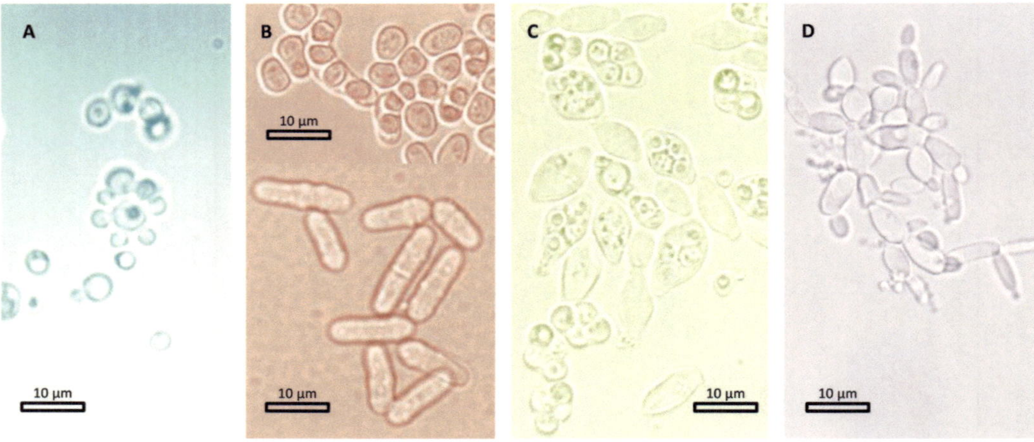

Figure 3. Optical microscopy of non-*Saccharomyces* yeasts: (A) *Torulaspora delbrueckii*, (B) *Schizosaccharomyces pombe*, (C) *Saccharomycodes ludwigii*, (D) *Dekkera* (*Brettanomyces*) *bruxelensis*.

spherical spores per ascus [22]. It has a low fermentative power reaching most of the strains a maximum of 6–9% (v/v) in ethanol. It has been described as osmotolerant [21] yeast with pure fermentation, with low production of volatile acidity and ethyl acetate.

It has also been described as a low producer of several volatile metabolites such as acetaldehyde, acetoin, and H_2S that can be unpleasant at high concentration [17, 25, 32]. Sequential fermentations with *S. cerevisiae* can be used to reduce volatile acidity [23]. *T. delbrueckii* also affects acidity balance by production of appreciable amounts of lactic and succinic acids [32]. Production of aromatic esters like ethyl lactate and 2-phenylethyl acetate is enhanced regarding single-culture *S. cerevisiae* fermentations. Some strains of *T. delbrueckii* have been described as stronger producers of 2-phenylethyl acetate in beers [33]. Ethyl lactate produces a coffee or strawberry smell that increases complexity also 2-phenylethyl acetate is the aromatic impact molecule of rose petals. 3-Ethoxy has been reported in sequential fermentations of *T. delbrueckii* and *S. cerevisiae* [24] with sensory repercussions for its either solvent of fruity (black currant) when combined with other flavors [34].

L. thermotolerans is getting attention in fermented beverages for its repercussion in sensory profile and the production of organic acids. Formerly it was known as *Kluyveromyces thermotolerans*. *L. thermotolerans* appearance is quite similar in both shape and size, to *S. cerevisiae* with an ellipsoidal morphology and impossible to be distinguish by optical microscopy. Asexual reproduction by multipolar budding also enables to sporulate with 1–4 spores in a dehiscent ascus. Its fermentation performance is medium and it can reach from 4 to 9% (v/v) in ethanol depending on the strain [25], but is unable to grow at higher contents than 9% [17]. Similar to *T. delbrueckii*, it shows good fermentation purity with low production of volatile acidity [35].

L. thermotolerans can be used to increase acidity during fermentation by production of lactic acid (**Table 1**). Concentrations of 9.6 g/L has been reported after fermentation [35].

http://dx.doi.org/10.5772/intechopen.68793

This acidification is enough to decrease pH significantly [25] and to affect sourness taste [36]. Moreover, the ability to ferment from malt sugars producing significant amounts of lactic acid make *L. thermotolerans* suitable to be used in the production of acidic beers without the involvement of lactic bacteria and in a single-stage fermentation [37].

Enhance the production of 2-phenylethanol and glycerol during fermentation [25, 36], *L. thermotolerans* has been described as an osmophilic yeast probably with higher production of glycerol, which is a tool to balance osmotic pressure. The levels of acetaldehyde and higher alcohols can be controlled using *L. thermotolerans* [38].

S. pombe is peculiar yeast traditionally used in some areas from Africa in the production of mijo beer. Pombe means beer in Swahili language. This yeast was isolated initially by Lindner on 1893 in East Africa. *S. pombe* has a rod-shaped structure with 3–4 μm in diameter and 7–20 μm in length. Asexual reproduction is performed by fission being a main difference with *S. cerevisiae*. It forms an intermediate septum at the center of the cell that is clearly visible by optical microscopy (**Figure 3B**). It is also teleomorph species having spherical spores in a linear organization with a typical amount of four per ascus (**Figure 3B**). The fermentative power is high being able to reach depending on the strain, 10–13% (v/v) in ethanol under anaerobiosis and 13–15 with slight aeration [39]. The production of acetic acid during fermentation is quite high (**Table 1**) being a main drawback in enology [20]; however, this parameter can be improved by selection, and also as the production depends on the amount of sugars metabolized in beer, the final levels are less conflictive. Other technological advantage is the resistance to sulfur dioxide.

Metabolism of organic acids is peculiar in *S. pombe* and different of its relative *S. cerevisiae*. *S. pombe* is a yeast that is able to degrade malic acid using the metabolic pathway call maloalcoholic fermentation which yields as main products ethanol and CO2. And the typical malic acid degradation done by lactic bacteria that also can be produced in beer or wine conditions uses other metabolic pathway call malolactic fermentation and produces lactic acid as main product. In the maloalcoholic fermentation, malate is decarboxylated to pyruvate by malic enzyme and later it is decarboxylated to acetaldehyde and finally reduced to ethanol [40]. Under anaerobiosis the fermentation of 2.3 g/L of malic acid produces 0.1% (v/v) of ethanol [41]. Degradation of malic acid can be higher than 8 g/L [39]. The sensory effect of this process is a softening of the sourness when substrates rich in malic acid are metabolized by *S. pombe*. The levels of pyruvate released during fermentation are also higher than the average of *S. cerevisiae* probably because of an intermediate in MA fermentation. The production of pyruvate affects the formation of pyranoanthocyanin pigments (vitisin A-type) during red wine fermentation by *S. pombe* [42]. Also, it has been reported that *S. pombe* can metabolize gluconic acid contents [43]. The unusual metabolism of organic acids in *S. pombe* can help to get other sourness balance in beers.

S. pombe is an osmophilic yeast with a double-layered cell wall composed of glucose, galactose, and mannose polysaccharides. Ageing on lees is a biological ageing in which fermented beverages remains together with the fermentation lees for a long time, more than 9 months. After yeast autolysis, cell wall polysaccharides are released and affect the sensory perception normally softening the mouth fell of the fermented beverages [27]. The thickness of the cell wall in *S. pombe* facilitates and increases the release of polysaccharides during ageing on lees.

Concentrations of 10 times higher in *S. pombe* regarding *S. cerevisiae* has been measured after 2 months of ageing of lees [27].

Some repercussions of *S. pombe* in food security of fermented beverages have been reported. The urease activity of *S. pombe* can facilitate the reduction of urea levels in musts reducing the risks of ethyl carbamate formation [39]. Also, the consumption of yeast assimilable nitrogen (YAN) by *S. pombe* is lower than in *S. cerevisiae*. So, it can be fermented substrates with low contents of YAN [44], minimizing the risk of production of nitrogen metabolites like biogenic amines.

S. ludwigii is an apiculated yeast lemon shaped with bipolar budding (**Figure 3C**). The cells are rather large in size 3–5 μm in diameter and 10–20 μm in length. It is a teleomorph yeast showing a typical rhomboid distribution with 4 spherical spores per ascus.

S. ludwigii has a strong resistance to sulfur dioxide due to the formation of electrophilic adducts with acetaldehyde [45]. The production and release of acetaldehyde during fermentation increases according to the concentration of free SO_2. Some strains produce high volatile acidity [46] but most of them are moderated; however, the production of ethyl acetate is frequently high (400 mg/L in average) [28].

The influence in aromatic profile is because of the production of esters giving a fruity taste. *S. ludwigii* expresses an extracellular β-glicosidase that can affect the release of free terpenes. Production of acetoin and diacetyl are also enhanced in *S. ludwigii* [47], many strains are able to produce more than 100 mg/L of acetoin and also some of them can reach 300 mg/L during fermentation [48].

Also, it has being described as an osmophilic yeast with thick cell wall and is able to release higher amounts of polysaccharides from the external covering during autolysis. The release of polysaccharides is quite similar to *S. pombe* and about 10-fold higher than *Saccharomyces* after 2 months of over lees ageing [27].

Traditionally, it is considered as spoilage yeast in enology because of the production of off-flavors, especially excessive amounts of acetoin and diacetyl. However, new applications are open currently in beer fermentation because of the lower amount of fermentable sugars. *S. ludwigii* is unable or a weak fermenter of either maltose or maltotriose from wort, and is being useful to produce alcohol-free and low-alcohol (0.5–1.2%, v/v ethanol) beers [8, 9, 49].

Dekkera bruxellensis and its anamorph form *B. bruxellensis* are ogival in shape with elongated cells that in old cultures can be highly branched because of the incomplete separation of the cells (**Figure 3D**). Its characteristics are multipolar budding, hat-shaped spores, film formation in liquid surface, and ethanol tolerance of 13–15% (v/v) [50].

The production of acetic acid by *Brettanomyces* is high (**Table 1**) and strongly dependent on the aeration conditions and oxygen availability [51]. Traditionally used in the production of sour beers (Belgian Lambics and Gueuze). Lambic beers are produced by a heterogeneous mixture of bacteria and yeasts in several phases. Acidification starts with the activity of lactic bacteria and even acetic bacteria, and is followed by the prevalence of *Brettanomyces,* mainly *bruxellensis* replacing the yeasts *Saccharomyces* at the end, when alcoholic degree is 5–6%

(v/v) [15]. *Brettanomyces* overferment the wort and degrades complex carbohydrates. At this time the population of lactic bacteria decreases and the flavor is enriched in the "Brett" taste.

In wines it has been described as producer of several off-flavors "Brett taint" [29, 52], most important are ethylpehnols. *Brettanomyces* is able to transform hydroxycinnamic acids in ethylphenols by mean of two enzymatic activities hydroxycinnamate decarboxylase that produce and intermediate vinylphenol and later vinylphenol reductase yielding the later ethylphenol. Descriptors of 4-ethylphenol are band-aid, leather, and horse sweat.

Brettanomyces release high amount of cell wall polysaccharides in ageing on lees [31] and could be an interesting parameter to modulate beer taste during bottle fermentation and ageing.

Selection, isolation, and counting of non-*Saccharomyces* can be done using selective media. Most non-*Saccharomyces* yeast can be separated of *S. cerevisiae* by culture in synthetic lysine phosphate media. *S. cerevisiae* can be differentiated from most of the non-*Saccharomyces* species because it is able to grow at 39°C [24].

4. Sensory effects of the use of non-*Saccharomyces* in beer conditioning and ageing

There are several hundreds of flavor-active compounds in beer. The main fermentation products of the brewing yeast are ethanol and carbon dioxide. Other molecules produced in smaller concentrations by yeasts during fermentation as metabolic intermediates or byproducts have great impact on beer flavor and determine the final quality of beer.

Although non-*Saccharomyces* yeasts have been widely disregarded due to their possible overproduction of acetic acid and other flavor compounds, they can potentially exert positive influences on beer flavor through the synthesis of secondary metabolites and excretion of enzymes responsible for the bioconversion of nonvolatile precursors into desirable aroma compounds. Using pure culture of a specific yeast species it is possible to modulate beer flavor by a natural biological method. These bioflavoring processes allows obtaining beer with enhanced and differentiated sensory profiles.

Several factors affect yeast fermentation with considerable influence on sensory profile of beer, some of the most relevant are pitching rate, temperature, duration, aeration, and C/N ratio [6, 15]. Also important is the specific species or yeast strain used [15]. Some of the most important descriptors in sensory analysis of beers are described in **Figure 4**.

Secondary metabolites can be divided in several groups including fusel alcohols, esters sulfur-containing flavor compounds, undesirable carbonyl compounds, volatile phenols, organic acids, and monoterpene alcohols [8]. Higher (fusel) alcohols and esters are the most important flavor-active substances in beer. Higher alcohols contribute to alcoholic taste, spicy, vinous, pungent aroma, and esters to fruity aroma of beer. Isoamyl acetate is considered a major contributor to the fruitiness of beer. However, it is possible that the presence of different esters below their threshold levels can exert a synergistic effect and play a role in beer flavor [53].

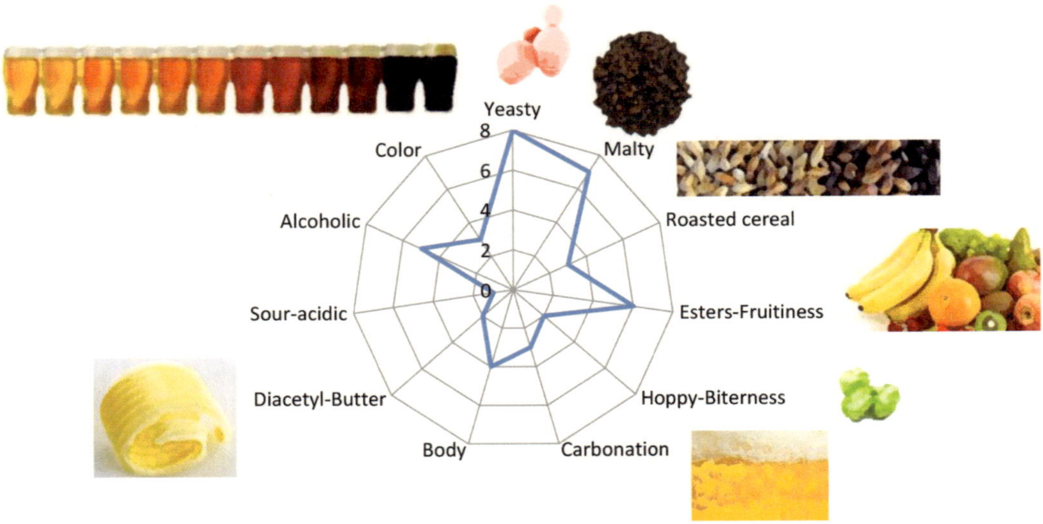

Figure 4. Spider net diagram for sensory profiling of beers.

The synthesis of fusel alcohols in beer fermentation is linked to the assimilation of the nitrogen sources by yeast and, therefore, typically the consumption and production of amino acids. Esters are formed later via reactions of alcohols (ethanol and fusels) and acids (AcylCoA compounds). These reactions are catalyzed by specific enzymes called acyl-alcohol transferases or esterases. When some non-*Saccharomyces* species fruity or floral ester production can be strongly enhanced, this biotechnology can be applied during main fermentation but also during second fermentation in the beers that are brewed and matured in bottle. *T. delbrueckii* and *L. thermotolerans* with medium fermentative power can be used to increase ester formation (**Table 1**).

Sulfur compounds are also produced by yeasts during fermentation, some of them behave as off-flavors and they have low sensory thresholds but others help to improve and modulate sensory profile of beers. Sulfur dioxide and hydrogen sulfide are included in these compounds [8], and they also have antioxidant properties and influence in flavor stability and increases shelf life. Hydrogen sulfide can be considered as a typical off-flavor in lager beers; however, at low concentrations it can be considered a typical aroma in some ale beers [54]. Generally, *Saccharomyces* strains used in ale fermentations are more prone to reduce sulfur dioxide to hydrogen sulfide than those used in lager beers.

Among carbonyl compounds the presence of diacetyl above the threshold levels is responsible of a buttery odor generally undesirable in lager beer but can be an interesting attribute in some beer styles such as English pale ales [53]. Diacetyl is reduced initially to acetoin and later to 2,3-butanediol, both with lower flavoring activity [54]. Malolactic fermentation by lactic acid bacteria (LAB) is a strategy widely used in wine making to enhance the production of diacetyl. In beer brewing, LABs are sensitive to hop bitter acids [53]. Increasing of temperature at the end of fermentation promotes the removal of diacetyl [54]. The levels of acetoin and diacetyl can also be enhanced by using non-*Saccharomyces* yeasts like *S. ludwigii* during main or alternatively bottle fermentation (**Table 1**). Yeast-produced phenols are responsible of desirable flavors with descriptors like clove, smoky, spicy, medicinal, and burnt aromas traditionally noticeable in wheat beer styles such as Belgian White beers, German Rauch beers, and Weizen beers. Typical non-*Saccharomyces* yeasts with strong formation of ethylphenols from hydroxycinnamates are *Brettanomyces* and *Dekkera* genera by their specific enzymatic activities (**Table 1**).

The main volatile organic acids that occur in beer are acetic, propionic, isobutyric, butyric, isovaleric, valeric, caproic, caprylic, capric, and lauric acid. If they are present at high concentrations they contribute to sour and salty flavor to beer and can also contribute to off-flavors such as cheesy and sweaty [8].

Lastly, some terpenic compounds from hops and with positive repercussion on beer flavor can be released and enhanced by some enzymes expressed by yeasts [8]. *Pichia anomala* and *Kloeckera apiculata* species frequently express β-glucosidase activities.

5. Sensory aspects related with engineering and brewing process

Beer-making process with inputs, byproducts, and residues can be seen in **Figure 5**. Wort boiling is the most energy-consuming operation and different alternatives to reduce the consumption of thermal energy have been implemented. High-gravity brewing has been a common practice in many commercial breweries, especially in lager beers [13]. Worts with a very high specific gravity (HSG; >16°P) are fermented to obtain a very high ethanol content beer. This beer is then diluted to reach the normal ethanol content.

According to Lodolo et al. [55] elevated osmotic pressure and the dilution of other essential nutritional factors such as amino acids contributes toward poor fermentation performance. The use of worts with high specific gravity (HSG) results in an unbalanced flavor profile [13, 14] and severe overproduction of acetate esters [13].

Another aspect related with brewing process is the use of large lager fermentation vessels in big-scale production. Fementor design can lead great pressures inside and excessive top pressure leads to poor yeast growth [13, 14]. The higher hydrostatic pressure in tall fermenters increases the concentration of carbon dioxide dissolved in beer and affects the stratification and irregular distribution of CO_2 during fermentation. The excess in dissolved CO_2 inhibits yeast growth and metabolism [13, 14]: poor diacetyl reduction and, most importantly, low ester production. However, according to Verstrepen et al. [13], in high gravity brewing (HGW) or when high fermentation temperatures are applied, the excessive formation of esters could be reduced by the use of tall fermenters.

On the other hand, some innovative brewer rooms are equipped by horizontal tanks that can be refrigerated by cooling jackets. Horizontal geometry leads to control yeast growth, modulates the ratio between higher alcohols and esters, and affects the flocculation, being possible to keep more suspended yeasts at the end of fermentation. Moreover, this geometry improves cleaning by CIP systems, and promotes the formation of aromatic esters.

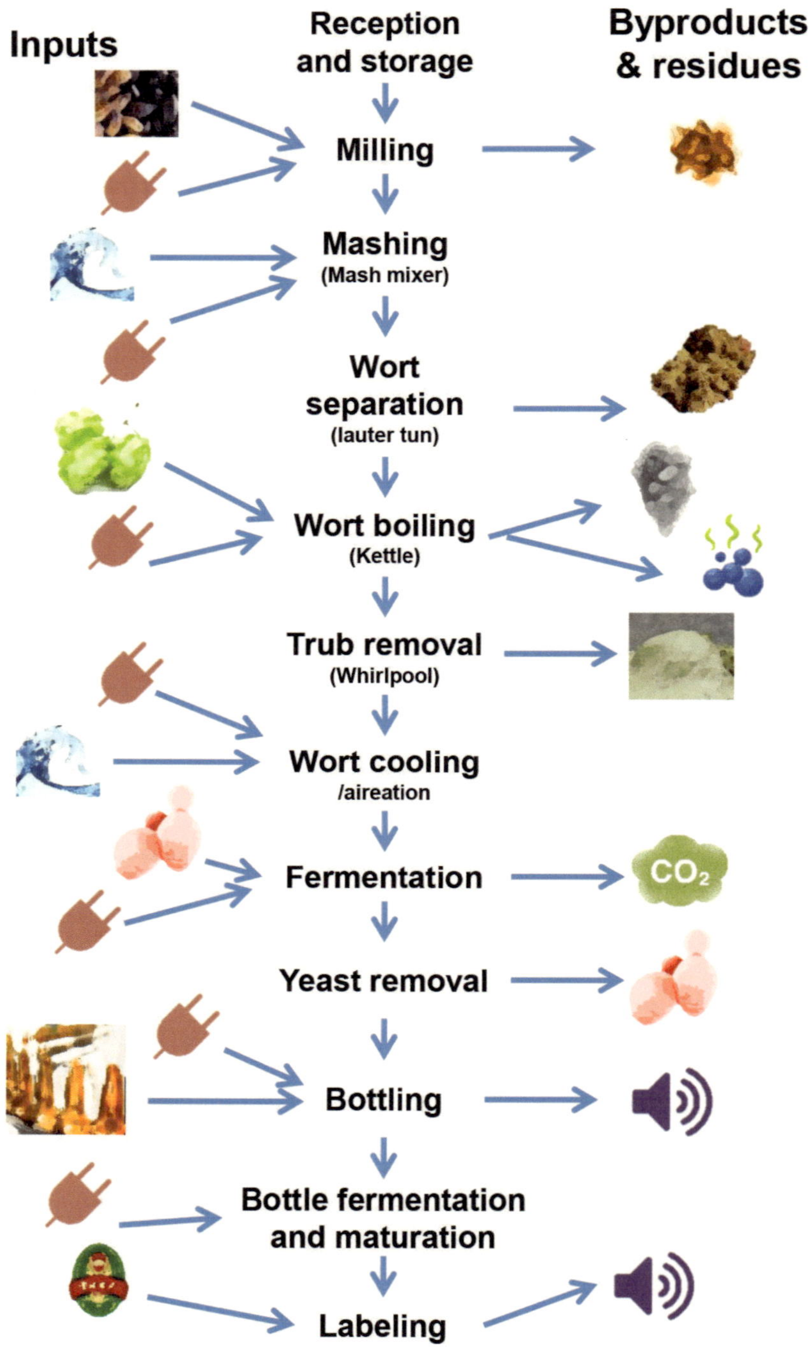

Figure 5. Inputs, byproducts, effluents, and residues in brewing process.

6. Conclusions and future trends

Non-*Saccharomyces* yeasts are a growing trend in new brewing biotechnology because of the improvements they can produce in sensory quality and differentiation, especially in craft beers. But also allows improved products as low-alcohol beers or implement new brewing processes. Probably in the future we will see new yeast species opening new possibilities of sensory and technological improvement. Bottle-conditioned beers are a good tool to obtain non-*Saccharomyces* bioflavored beers. Inoculation of new yeast species on a base beer completely attenuated with the addition of priming sugars allows creating optimal conditions to these yeasts.

Author details

María Jesús Callejo, Carmen González and Antonio Morata*

*Address all correspondence to: antonio.morata@upm.es

Department of Chemistry and Food Technology, Technical University of Madrid, Madrid, Spain

References

[1] Hornsey I. The Chemistry and Biology of Winemaking. Cambridge: The Royal Society of Chemistry; 2007. ISBN 9780854042661

[2] Anderson RG. History of Industrial Brewing. Food Science and Technology. Vol. 1. New York: Marcel Dekker; 2006. p. 157

[3] Pasteur L. Études sur la Bière. Paris: Gauthier-Villars; 1876

[4] Basso RF, Alcarde AR, Portugal CB. Could non-Saccharomyces yeasts contribute on innovative brewing fermentations? Food Research International. 2016;**86**:112-120. http://dx.doi.org/10.1016/j.foodres.2016.06.002

[5] Brewers-Association. National Beer Sales and Production Data. The New Brewer [Internet]. 2016. Available from: https://www.brewersassociation.org/statistics/national-beer-sales-production-data [Accessed: 21 March, 2016]

[6] Bokulich NA, Bamforth CW. The microbiology of malting and brewing. Microbiology and Molecular Biology Reviews. 2013;**77**:157-172. http://dx.doi.org/10.1128/MMBR.00060-12

[7] Steensels J, Verstrepen KJ. Taming wild yeast: Potential of conventional and nonconventional yeasts in industrial fermentations. Annual Review of Microbiology. 2014;**68**:61-80. http://dx.doi.org/10.1146/annurev-micro-091213-113025

[8] Michel M, Meier-Dörnberg T, Fritz J, Methner F-J, Wagner RS, Hutzler M. Review: Pure non-Saccharomyces starter cultures for beer fermentation with a focus on secondary

metabolites and practical applications. Journal of the Institute of Brewing. 2016;**122**:569-587. http://dx.doi.org/10.1002/jib.381

[9] De Francesco G, Turchetti B, Sileoni V, Marconi O, Perretti G. Screening of new strains of Saccharomycodes ludwigii and Zygosaccharomyces rouxii to produce low-alcohol beer. Journal of the Institute of Brewing. 2015;**121**:113-121. http://dx.doi.org/10.1002/jib.185

[10] Montanari L, Marconi O, Mayer H, Fantozzi P. Production of alcohol-free beer. In: Preedy VR, editor. Beer in Health and Disease Prevention. Elsevier: Amsterdam, The Netherlands, 2009; pp. 61-75

[11] Brányik T, Silva DP, Baszczyňski M, Lehnert R, Silva JBA. A review of methods of low alcohol and alcohol-free beer production. Journal of Food Engineering. 2012;**108**:493-506. http://dx.doi.org/10.1016/j.jfoodeng.2011.09.020

[12] Vanderhaegen B, Neven H, Verachtert H, Derdelinckx G. The chemistry of beer aging—A critical review. Food Chemistry. 2006;**95**:357-381. http://dx.doi.org/10.1016/j.foodchem.2005.01.006

[13] Verstrepen KJ, Derdelinckx G, Dufour JP, Winderickx J, Thevelein JM, Pretorius IS, Delvaux FR. Flavor-active esters: Adding fruitiness to beer. Journal of Bioscience and Bioengineering. 2003;**96**:110-118. http://dx.doi.org/10.1016/S1389-1723(03)90112-5

[14] Pires EJ, Teixeira JA, Brányik T, Vicente AA. Yeast: The soul of beer's aroma—A review of flavour-active esters and higher alcohols produced by the brewing yeast. Applied Microbiology and Biotechnology. 2014;**98**:1937-1949. http://dx.doi.org/10.1007/s00253-013-5470-0

[15] Crauwels S, Steensels J, Aerts G, Willems KA, Verstrepen KJ, Lievens B. Brettanomyces Bruxellensis, essential contributor in spontaneous beer fermentations providing novel opportunities for the brewing industry. Brewing Science. 2015;**68**:110-121

[16] Suárez-Lepe JA, Morata A. New trends in yeast selection for winemaking. Trends in Food Science & Technology. 2012;**23**:39-50. http://dx.doi.org/10.1016/j.tifs.2011.08.005

[17] Morata A, Suárez-Lepe JA. New biotechnologies for wine fermentation and ageing. In: Ravishankar Rai V, editor. Advances in Food Biotechnology. 1st ed. West Sussex, UK: John Wiley & Sons, Ltd; 2016. pp. 287-301

[18] Vaughan-Martini A, Martini A. Saccharomyces Meyen ex Reess (1870). In: Kurtzman C, Fell JW, Boekhout T, editors. The Yeasts: A Taxonomic Study. Chapter 61. Part IVB. Description of Teleomorphic Ascomycetous Genera and Species. Elsevier, Amsterdam, The Netherlands; 2011. pp. 737-738

[19] Snauwaert I, Roels SP, Van Nieuwerburg F, Van Landschoot A, De Vuyst L, Vandamme P. Microbial diversity and metabolite composition of Belgian red-brown acidic ales. International Journal of Food Microbiology. 2016;**221**:1-11. http://dx.doi.org/10.1016/j.ijfoodmicro.2015.12.009

[20] Loira I, Morata A, Comuzzo P, González MC, Calderón F, Suárez-Lepe JA. Use of Schizosaccharomyces pombe and Torulaspora delbrueckii strains in mixed and sequential fermentations for improving red wine sensory quality. Food Research International. 2015;**76**:325-333. http://dx.doi.org/10.1016/j.foodres.2015.06.030

[21] Hernandez-Lopez M, Prieto J, Randez-Gil F. Osmotolerance and leavening ability in sweet and frozen sweet dough. Comparative analysis between Torulaspora delbrueckii and Saccharomyces cerevisiae baker's yeast strains. Antonie Van Leeuwenhoek. 2003;**84**:125-134. http://dx.doi.org/10.1023/A:1025413520192

[22] Barnett JA. A history of research on yeasts 8: Taxonomy. Review. Yeast. 2004;**21**:1141-1193. http://dx.doi.org/10.1002/yea.1154

[23] Ciani M, Beco L, Comitini F. Fermentation behaviour and metabolic interactions of multistarter wine yeast fermentations. International Journal of Food Microbiology. 2006;**108**:239-245. http://dx.doi.org/10.1016/j.ijfoodmicro.2005.11.012

[24] Loira I, Vejarano R, Bañuelos MA, Morata A, Tesfaye W, Uthurry C, Villa A, Cintora I, Suárez-Lepe JA. Influence of sequential fermentations with Torulaspora delbrueckii and Saccharomyces cerevisiae on wine quality. LWT—Food Science and Technology. 2014;**59**:915-922. http://dx.doi.org/10.1016/j.lwt.2014.06.019

[25] Comitini F, Gobbi M, Domizio P, Romani C, Lencioni L, Mannazzu I, Ciani M. Selected non-Saccharomyces wine yeasts in controlled multistarter fermentations with Saccharomyces cerevisiae. Food Microbiology. 2011;**28**:873-882. http://dx.doi.org/10.1016/j.fm.2010.12.001

[26] Vaughan-Martini A, Martini A. Schizosaccharomyces lindner. In: Kurtzman C, Fell JW, editors. The Yeasts-A Taxonomic Study. Elsevier, Amsterdam, The Netherlands; 1998. p. 393

[27] Palomero F, Morata A, Benito S, Calderón F, Suárez-Lepe JA. New genera of yeasts for over-lees aging of red wine. Food Chemistry. 2009;**112**:432-441. http://dx.doi.org/10.1016/j.foodchem.2008.05.098

[28] Domizio P, Romani C, Lencioni L, Comitini F, Gobbi M, Mannazzu I, Ciani M. Outlining a future for non-Saccharomyces yeasts: Selection of putative spoilage wine strains to be used in association with Saccharomyces cerevisiae for grape juice fermentation. International Journal of Food Microbiology. 2011;**147**:170-180. http://dx.doi.org/10.1016/j.ijfoodmicro.2011.03.020

[29] Suárez R, Suárez-Lepe JA, Morata A, Calderón F. The production of ethylphenols in wine by yeasts of the genera Brettanomyces and Dekkera: A review. Food Chemistry. 2007;**102**:10-21. http://dx.doi.org/10.1016/j.foodchem.2006.03.030

[30] Blomqvist J, Eberhard T, Schnürer J, Passoth V. Fermentation characteristics of Dekkera bruxellensis strains. Applied Microbiology and Biotechnology. 2010;**87**:1487-1497. http://dx.doi.org/10.1007/s00253-010-2619-y

[31] Kulkarni P, Loira I, Morata A, Tesfaye W, González MC, Suárez-Lepe JA. Use of non-Saccharomyces yeast strains coupled with ultrasound treatment as a novel technique to accelerate ageing on lees of red wines and its repercussion in sensorial parameters. LWT—Food Science and Technology. 2015;**64**:1255-1262. http://dx.doi.org/10.1016/j.lwt.2015.07.046

[32] Ciani M, Maccarelli F. Oenological properties of non-Saccharomyces yeasts associated with wine-making. World Journal of Microbiology & Biotechnology. 1998;**14**:199-203

[33] Michel M, Kopecká J,Meier-Dörnberg T, Zarnkow M, Jacob F, Hutzler M. Screening for new brewing yeasts in the non-Saccharomyces sector with Torulaspora delbrueckii as model. Yeast. 2016;**33**:129-144. http://dx.doi.org/10.1002/yea.3146

[34] Tao Y, Zhang L. Intensity prediction of typical aroma characters of cabernet sauvignon wine in Changli County (China). LWT—Food Science and Technology. 2010;**43**:1550-1556. http://dx.doi.org/10.1016/j.lwt.2010.06.003

[35] Kapsopoulou K, Kapaklis A, Spyropoulos H. Growth and fermentation characteristics of a strain of the wine yeast Kluyveromyces thermotolerans isolated in Greece. World Journal of Microbiology and Biotechnology. 2005;**21**:1599-1602. http://dx.doi.org/10.1007/s11274-005-8220-3

[36] Gobbi M, Comitini F, Domizio P, Romani C, Lencioni L, Mannazzu I, Ciani M. Lachancea thermotolerans and Saccharomyces cerevisiae in simultaneous and sequential co-fermentation: A strategy to enhance acidity and improve the overall quality of wine. Food Microbiology. 2013;**33**:271-281. http://dx.doi.org/10.1016/j.fm.2012.10.004

[37] Domizio P, House JF, Joseph CML, Bisson LF, Bamforth CW. Lachancea thermotolerans as an alternative yeast for the production of beer. Journal of the Institute of Brewing. 2016;**122**:599-604. http://dx.doi.org/10.1002/jib.362

[38] Balikci EK, Tanguler H, Jolly NP, Erten H. Influence of Lachancea thermotolerans on cv. Emir wine fermentation. Yeast. 2016;**33**:313-321. http://dx.doi.org/10.1002/yea.3166

[39] Suárez-Lepe JA, Palomero F, Benito S, Calderón F, Morata A. Oenological versatility of Schizosaccharomyces spp. European Food Research Technology. 2012;**235**:375-383. http://dx.doi.org/10.1007/s00217-012-1785-9

[40] Maconi E, Manachini PL, Aragozzini F, Gennari C, Ricca GS. A study of the maloalcoholic fermentation pathway in Schizosaccharomyces pombe. Biochemical Journal. 1984; **217**:585-588

[41] Taillandier P, Strehaiano P. The role of malic acid in the metabolism of Schizosaccharomyces pombe: Substrate consumption and cell growth. Applied Microbiology and Biotechnology. 1991;**35**:541-543. http://dx.doi.org/10.1007/BF00169765

[42] Morata A, Benito S, Loira I, Palomero F, González MC, Suárez-Lepe JA. Formation of pyranoanthocyanins by Schizosaccharomyces pombe during the fermentation of red must. International Journal of Food Microbiology. 2012;**159**:47-53. http://dx.doi.org/10.1016/j.ijfoodmicro.2012.08.007

[43] Peinado RA, Mauricio JC, Medina M, Moreno JJ. Effect of Schizosaccharomyces pombe on aromatic compounds in dry sherry wines containing high levels of gluconic acid. Journal of Agricultural and Food Chemistry. 2004;**52**:4529-4534. http://dx.doi.org/10.1021/jf049853r

[44] Benito S, Palomero F, Morata A, Calderón F, Suárez-Lepe JA. New applications for Schizosaccharomyces pombe in the alcoholic fermentation of red wines. International Journal of Food Science & Technology. 2012;**47**:2101-2108. http://dx.doi.org/10.1111/j.1365-2621.2012.03076.x

[45] Stratford M, Morgans P, Rose AH. Sulphur dioxide resistance in Saccharomyces cerevisiae and Saccharomycodes ludwigii. Journal of General Microbiology. 1987;**133**:2173-2179

[46] Romano P, Marchese R, Laurita C, Saleano G, Turbanti L. Biotechnological suitability of Saccharomycodes ludwigii for fermented beverages. World Journal of Microbiology and Biotechnology. 1999;**15**:451-454. http://dx.doi.org/10.1023/A:1008948623024

[47] Escott C, Morata A, Loira I, Tesfaye W, Suarez-Lepe JA. Characterization of polymeric pigments and pyranoanthocyanins formed in microfermentations of non-Saccharomyces yeasts. Journal of Applied Microbiology. 2016;**121**:1346-1356. http://dx.doi.org/10.1111/jam.13255

[48] Romano P, Suzzi G. Origin and production of acetoin during wine yeast fermentation. Applied and Environmental Microbiology. 1996;**62**:309-315

[49] Liu, Y., Li, H., & Du, J. H. (2011). Non-alcoholic beer production by Saccharomycodes ludwigii. Food Science, 32, 186-190.

[50] Barata A, Caldeira J, Botelheiro R, Pagliara A, Malfeito-Ferreira M, Loureiro V. Survival patterns of Dekkera bruxellensis in wines and inhibitory effect of Sulphur dioxide. International Journal of Food Microbiology. 2008;**121**:201-207. http://dx.doi.org/10.1016/j.ijfoodmicro.2007.11.020

[51] Aguilar-Uscanga MG, Délia ML, Strehaiano P. Brettanomyces bruxellensis: Effect of oxygen on growth and acetic acid production. Applied Microbiology and Biotechnology. 2003;**61**:157-162. http://dx.doi.org/10.1007/s00253-002-1197-z

[52] Loureiro V, Malfeito-Ferreira M. Spoilage yeasts in the wine industry. International Journal of Food Microbiology. 2003;**86**:23-50. http://dx.doi.org/10.1016/S0168-1605(03)00246-0

[53] Yeo HQ, Liu SQ. An overview of selected specialty beers: Developments, challenges and prospects. International Journal of Food Science & Technology. 2014;**49**:1607-1618. http://dx.doi.org/10.1111/ijfs.12488

[54] Baxter ED, Hughes PS. Cerveza: Calidad, Higiene y Características Nutricionales. Zaragoza: Acribia; 2004

[55] Lodolo EJ, Kock JL, Axcell BC, Brooks M. The yeast Saccharomyces cerevisiae—The main character in beer brewing. FEMS Yeast Research. 2008;**8**(7):1018-1036. http://dx.doi.org/10.1111/j.1567-1364.2008.00433.x

Section 3

Brewing Technology

Chapter 6

Concept of *Nuruk* on Brewing Technology

Jang-Eun Lee and Jae-Ho Kim

Additional information is available at the end of the chapter

http://dx.doi.org/10.5772/intechopen.69380

Abstract

Nuruk is a traditional Korean fermentation starter that is used to produce starch-based alcoholic beverages using various cereals as raw material. As a determinant factor for flavor, taste, and color of alcoholic beverages, *Nuruk* is an indispensable ingredient for brewing alcoholic beverages in Korea. *Nuruk* shows significant variation in the shape, and in the brewing and fermentation methods, which are dependent on the unique climate in each area. Therefore, it is worthy to note that the characteristics of Korean traditional *Nuruk* are based on its diversity. Thus, this chapter is aimed to scientifically identify the characteristics of traditional *Nuruk* on brewing technology. In this chapter, the concept of *Nuruk* will be discussed in terms of its history, production, microorganism diversity, and enzymatic function.

Keywords: fungi, microorganisms, *Nuruk*, saccharification power, yeast

1. Introduction

Nuruk, a fermentation starter used for brewing alcoholic beverages from grains, is a dough made from grains, such as wheat, barley, or rice that are germinated by enzyme-releasing microorganisms. In Asian countries, starch is the main ingredient for alcohol fermentation, which is first hydrolyzed to glucose through the saccharification process by fungi. *Nuruk*, also called "Gokja" in Korea, contains naturally occurring and multiplying microorganisms such as wild fungi, yeast, and lactic acid bacteria. Traditionally, *Nuruk* has been made from several grains such as wheat, barley, rice, and millet etc., and grains are used as the main raw material for *Nuruk* and alcohol beverages processing.

In Korea, *Nuruk* shows significant variation in the shape, and in the brewing and fermentation methods, which are dependent on the unique climate in each area. It has been shown that *Nuruk* production can be adapted to suit the geographical area and climate. With the development of molded *Nuruk* in China and the dispersed *Koji* in Japan, Korean traditional *Nuruk* has been developed with a wide variety of materials and shapes. For its development, the humidity and the amount of sunshine strongly influences the width and thickness of *Nuruk* originating from different areas in Korea from the wide and thin *Nuruk* in the mountainous areas, to the thick and small *Nuruk* in the flat areas. In addition, the main raw materials used to produce *Nuruk* vary widely according to the production area. Therefore, it is worthy to note that the characteristics of Korean traditional *Nuruk* depend on the geographical and climatic diversities of their production area. However, the use of traditional *Nuruk*, in a variety of traditional brewing method is dwindling due to the recent increase in the use of industrial commercial fermentation starter [1]. In Korea, research on traditional *Nuruk* was started in the early 1900s by the Japanese, but had not actively progressed until recently. Currently, research on the production of certain traditional *Nuruk* [2–4], *Nuruk*-derived microorganisms [5–9], traditional *Nuruk*-derived Korean alcoholic beverages [10–13], and their physiological functions [14–17] are being performed. In addition, investigation on the *Nuruk* microbial communities [18–20] as well as the metabolite analysis of *Nuruk* [20] was recently performed.

In this chapter, the use of *Nuruk* in brewing technology will be discussed in terms of its history, production, microorganism diversity, and enzyme function.

2. History of *Nuruk*

Nuruk was first made in Asia in the fifth century BC. It is believed that *Nuruk* was first used in Korea before the "three kingdoms period," and records show *Nuruk* being used for Korean alcohol production in the 1123 CE book *Goryeo Dogyeong* (Chi.: *Gaolitujing*) by Xu Jing. *Hallimbyeolgok* from the Goryeo period mentions an alcohol brewed with a special type of *Nuruk*, indicating the existence of several types of *Nuruk* in Korea at that time. *Gyugonsiuibang* (1670), a classic text about food in the mid-Joseon period, records the names and detailed manufacturing methods for different types of *Nuruk*, highlighting that a diverse range of traditional *Nuruk*s were manufactured during that period. In classic texts, *Nuruk* was called Guk, and after 1918, it was called Gokja; however, currently the term "*Nuruk*" is more common than Gokja.

During the reign of the Joseon dynasty, *Nuruk* was classified into two categories: the *ddeok-Nuruk* was made of a lump of grain powder and the *Heuchim-Nuruk* was made of cereal grains. The appearances of *ddeok*-and *Heuchim-Nuruk* are presented in **Figure 1**. The *ddeok-Nuruk* contained a variety of microorganisms such as fungi, lactic acid bacteria, and yeasts deep inside the lump, which imparted rich and complex flavor to alcohol. In contrast, fungi germinating only on the surface of the *heuchim-Nuruk*, provided simple and light tastes.

The dry climate of China favored the development of the shaped *Nuruk* from wheat, whereas the humidity of the Japanese climate promoted the development of dispersed *Nuruk* from rice. Meanwhile, Korea developed both the types of *Nuruk*. Thus, *Nuruk*s differ in particle shape, manufacturing methods, and fermentation time, depending on the unique climate and environment of the manufacturing country, thereby exemplifying the adaptation ability of

Nuruk to climatic and geographical conditions. In contrast to the shaped and dispersed *Nuruk* developed by China and Japan, respectively, Korea developed *Nuruk*s with diverse ingredients and appearances. For example, *Nuruk* grain particles of mountainous regions tend to be broad and flat, whereas *Nuruk* particles of the plains are thick and small, and those in the Jeju region are small and flat. These regional differences occur due to variations in the content of the main ingredients and the environment, such as the levels of humidity and sunlight. Several kinds of representative traditional Korean *Nuruk* are displayed in **Figure 2**. As shown in **Figure 2**, this diversity is the characteristic of the traditional Korean *Nuruk*, which has been promoted by the development of a unique and varied traditional home brewing culture.

Figure 1. Dispersed *heuchim-Nuruk* (A) and shaped *ddeok-Nuruk* (B and C).

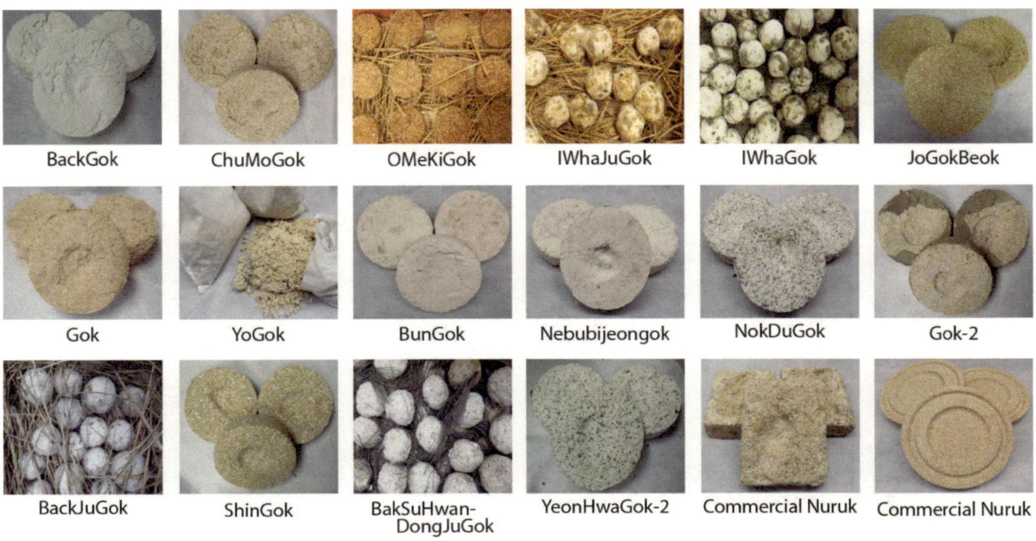

Figure 2. Appearance of various traditional Korean *Nuruk* [21].

3. *Nuruk* production

To ferment *Nuruk*, fungi or bacteria are germinated on a culture medium which is made of starchy grains such as rice, wheat, and barley. Wheat and barley are the most popular materials for *Nuruk* fermentation as they impart quality taste and flavor to *Nuruk*.

The *Nuruk* manufacture method is summarized in **Figure 3**. Traditionally, ground wheat is mixed with water, put in a mold, and pressed into the desired shape (**Figure 4**). Whole grains are thoroughly ground and finely sifted, mixed with other supplemental materials, and pressed into a frame to shape *Nuruk*. The shaped *Nuruk* is germinated with microorganisms for 2–3 days buried under supplementary materials such as straw or wormwood at a temperature of 30–35°C. The growth of yellowish fungi in the center of the pressed mass indicates that the *Nuruk* should be dried under the sun, thoroughly crushed, and finely sifted. Favorable temperature and humidity are critical to the culture of fungi on *Nuruk*. *Nuruk* can be globe-shaped, flat round disk-shaped, or rectangular with a hole in the center. *Nuruk* must be made in just the right size and thickness. Small and thin *Nuruk* loses moisture easily, which causes incomplete germination of fungi and defective fermentation, resulting in undesirable flavor, and low alcohol yield. In contrast, a thick *Nuruk* limits moisture loss and increases the temperature inside the fermentation jar. A well-cultured *Nuruk* is critical for the clear color and fresh flavor of fermented grain.

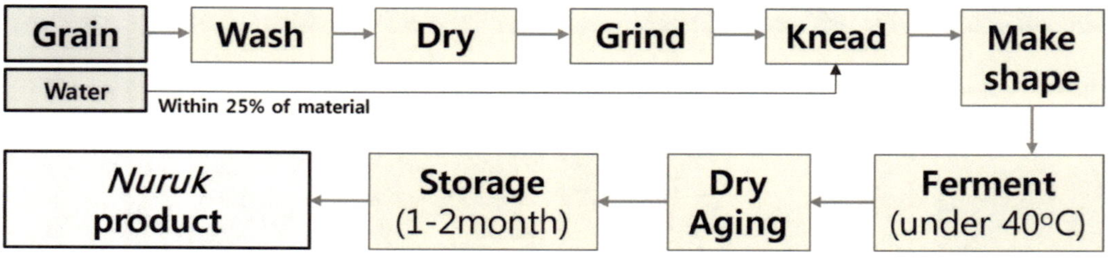

Figure 3. *Nuruk* production.

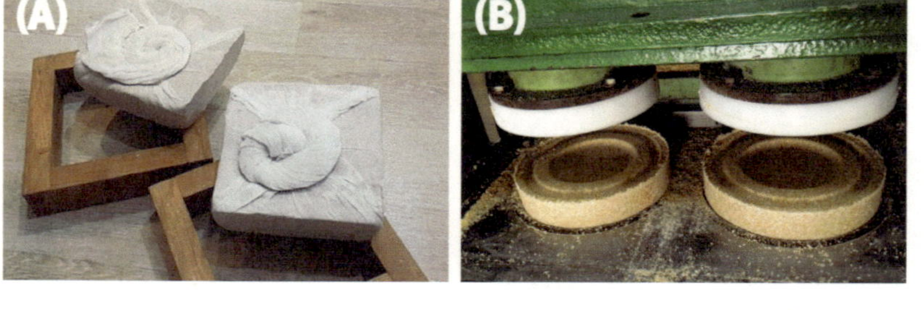

Figure 4. *Nuruk* is put in a mold and pressed into the desired shape. (A) traditional molding method, (B) mechanization for mass produce Nuruks.

4. Microorganisms in *Nuruk*

Various types of microbes exist in *Nuruk* because of the coexistence of raw material (raw starch)-derived microorganisms and environment-derived microorganisms that were acquired during the fermentation. Since *Nuruk* mainly consists of starch, microorganisms that are capable of degrading beta-starch are predominant in *Nuruk*. Specifically, various types of yeast, lactic acid bacteria, and aerobic bacteria that cause alcoholic fermentation are present in *Nuruk*. Some kinds of *Nuruk* derived microorganisms that were found by Kim et al. [14] are shown in **Figure 5**.

In general, the fungus that grows in yeast has high starch decomposition activity, and thus hydrolysis of starch during alcoholic fermentation produces fermentable saccharides. In addition, fungi also produce alcohol from fermentable sugars. Lactic acid bacteria are involved in maintaining the acidity of the fermentation environment, which enables progression of alcoholic fermentation by acid-producing fungi. Mold, a group of mesophilic fungi that grows well at 25–30°C and in weakly acidic conditions, plays an important role in brewing because it secretes glycosylation enzymes required for hydrolyzing starch into fermentable saccharides. Therefore, fermentative molds in low pH environment exhibit high liquefaction and glycation

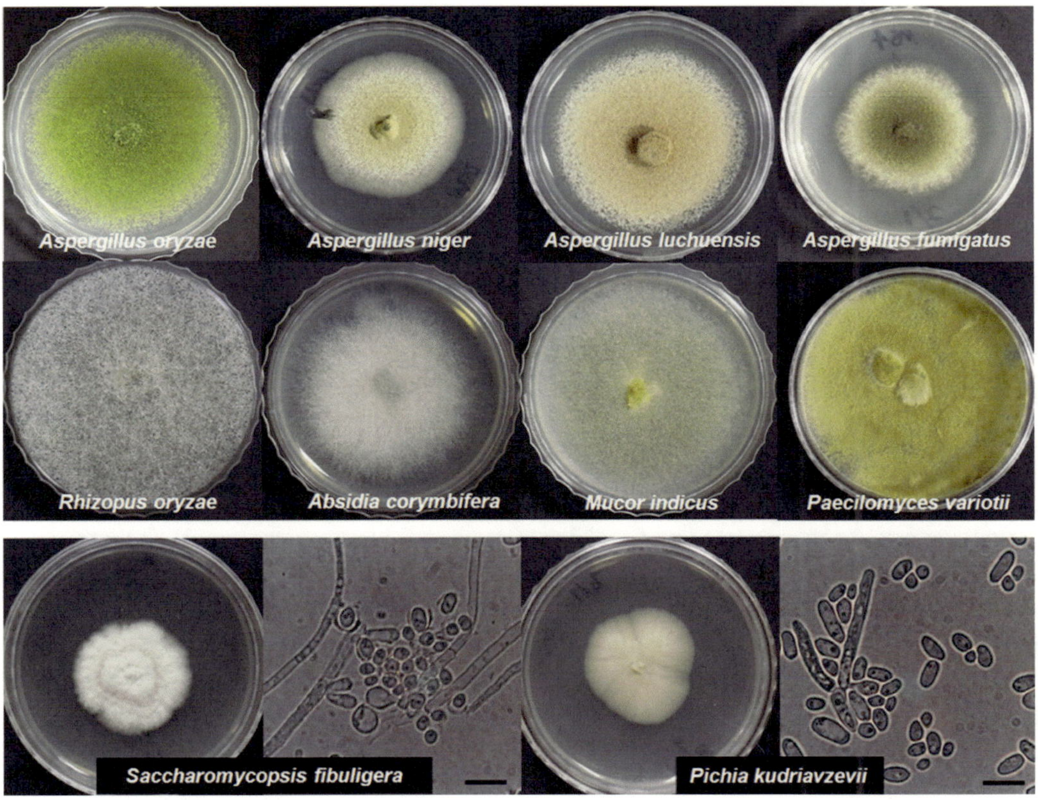

Figure 5. The filamentous fungi and yeasts isolated from various *Nuruk* samples (scale bar = 5 μm) [14].

abilities. The most common molds found in *Nuruk* are *Absidia* sp, *Aspergillus oryzae, Rhizopus* sp., *Penicillium* sp., and *Muco* sp., which secrete various amylases such as α-amylase and glucoamylase that act on stored starch of cereals. The molds most frequently isolated from *Nuruk* are *R. oryzae* and *A. oryzae*. In Asia, these two strains are important fungi in the food industry because they are used in the manufacture of traditional fermented foods. *R. oryzae* is a pathogenic fungus that causes zygomycosis [22] and, is a pioneer saprotroph that initially infects dead plants and rapidly penetrates inside degradable substrates. Recent genomic studies showed that *R. oryzae* does not contain the genes necessary for exo-cellulose degradation and is an auxotroph for degradation of pectin, xyloglucan, xylan, and inulin. Thus, the ability to degrade simple polysaccharides, such as monosaccharides and starches, underscores the dominance of *R. oryzae* over other fungi [23].

Among yeasts, *R. oryzae* acts as a primary colonizer in the early stages of fermentation and participates in the decomposition of yeast constituents. However, as the internal temperature of the mold increases with progression of fermentation, the temperature-sensitive *R. oryzae* gradually falls out of competition with the high temperature-resistant fungi such as *A. oryzae* [24]. *A. oryzae* possesses high glycation ability and secretes α-amylase [25], and is therefore important for glycosylation and liquefaction [8, 24]. However, it is difficult to distinguish *A. oryzae* from aflatoxin-producing *A. flavus* using conventional taxonomic criteria because of its close relationship with *A. flavus*; thus, *A. oryzae* should be tested for its ability to produce aflatoxin [26, 27]. The members of the aflatoxin biosynthetic pathway are encoded by a cluster of 25 genes [28], the expression levels of which are primarily regulated by aflR, a transcriptional activator [29]. Therefore, polymorphism, deletion, and the presence of aflR binding sites have been used as important indicators to determine aflatoxin biosynthesis [26]. Particularly, the aflR binding sites of *norB* and *cypA* are critical for the expression of the entire aflatoxin biosynthesis-associated gene group. In general, *A. oryzae* and *A. flavus* are divided into type I deletion (0.3 kb) and type II deletion (0.8 kb) groups based on the partial deletion pattern of *norB-cypA* (1.8 kb), and it was found that the type I deletion group does not produce aflatoxin [27, 30, 31].

5. Enzymatic activity of *Nuruk*

There are a variety of microorganisms in fermented *Nuruk* and enzymes secreted by these microorganisms differ depending on the type of wild microorganism. Enzymatic activity of *Nuruk* can mainly be estimated by the index of saccharification power (sp). This saccharification power represents the index of how much soluble starch can be enzymatically converted to simpler sugars by diastatic enzymes in *Nuruk*. Previous studies on the enzymatic activity of early *Nuruk* showed that the saccharification power (sp) of Korean *Nuruk* was 1.39, which was slightly lower than the saccharification power of malt (1.5 sp), and significantly lower than that of Chinese *Nuruk* (11.1 sp) [32]. According to previous study [33], it has been reported that the saccharification power of *Backguk* produced by *Aspergillus kawachii and Hwangguk* produced by *Asp. oryzae* were being used in all areas of Korea in the mid-1900s. The saccharification power values were the highest in *Bungok* (791 sp), followed by *Gokja* (421 sp), *Hwangguk* (226 sp), and *Backguk* (195 sp).

No.	*Nuruk*	Saccharification power	Enzyme activity[a]				
			α-Amylase (U/g *Nuruk*)	β-Amylase (U/g *Nuruk*)	Protease (μg/ml trypsin)	β-Glucanase (U/g *Nuruk*)	Glucoamylase (U/g *Nuruk*)
1	JuKokBangMun	325.43 ± 50.97	1.58 ± 0.36	2.82 ± 0.17	1.53 ± 0.09	0.048 ± 0.008	0.16 ± 0.10
2	ChuMoGok	194.82 ± 0.10	17.42 ± 1.10	1.44 ± 0.76	1.57 ± 0.07	0.191 ± 0.020	0.17 ± 0.06
3	OMeKiGok	151.44 ± 16.11	17.84 ± 0.86	0.72 ± 0.13	9.90 ± 7.28	0.348 ± 0.022	0.48 ± 0.29
4	JukGok-1	319.34 ± 76.83	10.75±1.81	3.13±0.25	1.77±0.41	0.042±0.001	0.75±0.58
5	JukGok-2	265.93 ± 18.50	6.72 ± 1.02	3.35 ± 0.52	1.61 ± 0.10	0.041 ± 0.004	0.43 ± 0.57
6	KongByungGok	318.83 ± 17.74	39.07 ± 6.70	4.58 ± 0.36	1.42 ± 0.87	0.108 ± 0.014	1.05 ± 0.09
7	Gok-1	346.75 ± 14.76	1.47 ± 0.43	3.83 ± 0.24	1.48 ± 0.36	0.046 ± 0.016	0.05 ± 0.02
8	MyunGok-1	292.22 ± 12.14	8.76 ± 1.77	4.37 ± 0.11	1.75 ± 0.22	0.057 ± 0.008	0.27 ± 0.06
9	SinGok-1	137.46 ± 10.23	6.0251.29	1.01 ± 0.31	2.31 ± 0.61	0.187 ± 0.017	0.16 ± 0.05
10	JinjuchunchuGok	163.53 ± 0.09	16.61 ± 2.85	0.18 ± 0.02	1.76 ± 0.31	0.091 ± 0.007	0.31 ± 0.06
11	YeoGok-1	85.53 ± 6.12	9.45 ± 6.73	0.05 ± 0.01	1.51 ± 0.25	0.082 ± 0.011	0.15 ± 0.03
12	SeolhangGok	211.52 ±1 8.21	2.73 ± 1.82	1.83 ± 0.30	1.63 ± 0.07	0.049 ± 0.007	0.04 ± 0.00
13	YeonHwaGok-1	112.67 ± 18.18	2.44 ± 0.57	0.05 ± 0.02	1.43 ± 0.13	0.161 ± 0.052	0.29 ± 0.08
14	BackGok-1	150.76 ± 11.76	11.47 ± 4.56	1.31 ± 0.07	1.59 ± 0.05	0.047 ± 0.008	0.04 ± 0.04
15	BackGok-1	164.61 ± 4.65	46.09 ± 10.57	1.56 ± 0.20	2.08 ± 0.09	0.047 ± 0.010	0.13 ± 0.01
16	YeoGok-2	370.09 ± 6.05	0.96 ± 0.05	2.74 ± 0.13	1.63 ± 0.08	0.043 ± 0.003	0.03 ± 0.01
17	BunGok	565.54 ± 22.67	65.53 ± 2.42	3.80 ± 0.34	1.82 ± 0.13	0.121 ± 0.043	0.50 ± 0.03
18	ByungGok	395.20 ± 47.49	53.20 ± 7.42	4.58 ± 0.19	1.61 ± 0.26	0.070 ± 0.013	0.37 ± 0.12
19	IWhaJuGok	241.07 ± 12.05	29.61 ± 3.60	0.09 ± 0.02	1.69 ± 0.30	0.334 ± 0.023	0.45 ± 0.06
20	IWhaGok	351.28 ± 10.74	19.90 ± 2.50	0.08 ± 0.00	2.09 ± 0.31	0.255 ± 0.014	0.46 ± 0.02
21	JoGokBeok	283.77 ± 17.23	2.33 ± 1.41	4.31 ± 0.02	1.61 ± 0.08	0.043 ± 0.010	0.10 ± 0.02

No.	*Nuruk*	Saccharification power	Enzyme activity[a]				
			α-Amylase (U/g *Nuruk*)	β-Amylase (U/g *Nuruk*)	Protease (μg/ml trypsin)	β-Glucanase (U/g *Nuruk*)	Glucoamylase (U/g *Nuruk*)
22	HyangOnGok-1	353.82 ± 35.81	0.69 ± 0.03	2.05 ± 0.19	1.58 ± 0.09	0.041 ± 0.009	0.04 ± 0.03
23	HyangOnGok-2	408.60 ± 42.88	0.78 ± 0.07	3.44 ± 0.56	1.62 ± 0.02	0.036 ± 0.004	0.03 ± 0.02
24	BakSuHwanDongJuGok	512.81 ± 6.26	28.09 ± 0.80	0.38 ± 0.07	2.64 ± 0.26	0.070 ± 0.011	1.07 ± 0.07
25	Nebubijeongok	565.20 ± 3.54	33.20 ± 2.24	4.11 ± 0.16	1.76 ± 0.29	0.050 ± 0.011	0.24 ± 0.03
26	NokMiJuGok	217.74 ± 24.29	21.17 ± 4.13	0.40 ± 0.05	3.33 ± 0.90	0.083 ± 0.032	1.09 ± 0.07
27	NokDuGok	210.59 ± 7.15	2.27 ± 1.01	0.10±0.01	1.78 ± 0.12	0.032 ± 0.003	0.17 ± 0.02
28	Gok-2	334.91 ± 21.23	16.42 ± 8.45	4.17 ± 0.26	2.75 ± 0.97	0.043 ± 0.005	0.34 ± 0.30
29	MiGok-1	547.90 ± 18.10	22.18 ± 6.78	4.77 ± 0.36	2.11 ± 0.33	0.041 ± 0.004	0.21 ± 0.07
30	MyunGok-2	265.45 ± 46.99	4.54 ± 0.73	3.78 ± 0.92	1.98 ± 0.20	0.053 ± 0.006	0.18 ± 0.04
31	Gok-3	190.28 ± 7.11	2.32 ± 2.10	4.27 ± 0.15	1.62 ± 0.12	0.042 ± 0.012	0.10 ± 0.07
32	YoGok	143.42 ± 10.75	1.33 ± 0.26	0.07 ± 0.02	1.59 ± 0.12	0.173 ± 0.085	0.04 ± 0.05
33	DaeJuBackTaGok	189.51 ± 0.59	3.62 ± 1.06	2.38 ± 0.07	1.66 ± 0.20	0.077 ± 0.008	0.14 ± 0.07
34	BackRyoGok	244.06 ± 2.56	4.23 ± 0.65	1.12 ± 0.07	1.70 ± 0.15	0.083 ± 0.021	0.07 ± 0.01
35	YangNeungGok	183.92 ± 35.08	0.71 ± 0.09	1.71 ± 0.07	1.64 ± 0.03	0.037 ± 0.004	0.01 ± 0.01
36	BackJuGok-1	214.55 ± 57.45	1.10 ± 0.07	0.04 ± 0.02	1.71 ± 0.27	0.282 ± 0.013	0.04 ± 0.01
37	BackJuGok-2	222.16 ± 20.73	2.69 ± 0.38	0.05 ± 0.02	1.43 ±0.15	0.046 ± 0.005	0.27 ± 0.05
38	ManJeonHangJuGok	267.99 ± 22.97	0.74 ± 0.03	2.84±0.19	1.65±0.04	0.037±0.003	0.02±0.02
39	JeongHwaGok	324.33 ± 5.46	43.41 ± 8.91	4.64 ± 0.62	2.16 ± 0.94	0.055 ± 0.003	0.47 ± 0.02
40	YeonHwaGok-2	516.06 ± 5.50	63.37 ± 5.13	0.68 ± 0.01	2.12 ± 0.15	0.046 ± 0.003	0.30 ± 0.02
41	DongYangJuGok	207.22 ± 87.05	23.14 ± 3.29	5.28 ± 0.36	2.11 ± 0.22	0.092 ± 0.006	0.21 ± 0.07
42	MiGok-2	101.99 ± 12.02	1.03 ± 0.12	0.05 ± 0.02	1.72 ± 0.06	0.043 ± 0.007	0.03 ± 0.01

No.	*Nuruk*	Saccharification power	Enzyme activity[a]				
			α-Amylase (U/g *Nuruk*)	β-Amylase (U/g *Nuruk*)	Protease (μg/ml trypsin)	β-Glucanase (U/g *Nuruk*)	Glucoamylase (U/g *Nuruk*)
43	ShinGok-1	356.61 ± 7.28	2.52 ± 0.63	0.30 ± 0.02	1.69 ± 0.33	0.038 ± 0.005	1.94 ± 0.17
44	ShinGok-2	340.37 ± 36.95	29.52 ± 4.28	4.14 ± 0.70	1.86 ± 0.20	0.174 ± 0.011	0.56 ± 0.12
45	ShinGok-3	215.02 ± 26.57	12.69 ± 0.80	1.79 ± 0.13	1.78 ± 0.22	0.050 ± 0.003	0.48 ± 0.09
46	HaDongShinGok	268.30 ± 8.59	4.29 ± 0.17	1.00 ± 0.06	1.88 ± 0.26	0.163 ± 0.013	0.07 ± 0.01
47	Commercial-1	423.59 ± 8.64	26.02 ± 5.22	4.23 ± 0.26	2.79 ± 1.58	0.150 ± 0.005	0.42 ± 0.08
48	Commercial-2	460.12 ± 18.81	30.33 ± 8.43	5.01 ± 0.47	5.04 ± 1.52	0.186 ± 0.025	0.53 ± 0.06
49	Commercial-3	356.56 ± 40.15	38.23 ± 17.56	5.18 ± 0.52	3.23 ± 2.60	0.133 ± 0.025	0.58 ± 0.33
50	Commercial-4	220.99 ± 23.81	21.98 ± 6.05	3.10 ± 0.20	1.70 ± 0.08	0.116 ± 0.012	0.75 ± 0.12
51	Commercial-5	418.80 ± 15.53	3.54 ± 0.65	0.90 ± 0.79	1.74 ± 0.11	0.047 ± 0.006	0.16 ± 0.06
52	Commercial-6	346.28 ± 57.71	3.46 ± 1.40	0.11 ± 0.03	1.72 ± 0.14	0.042 ± 0.006	0.13 ± 0.06
53	Commercial-7	229.63 ± 14.59	8.01 ± 1.27	1.98 ± 0.06	1.72 ± 0.09	0.083 ± 0.002	0.27 ± 0.02
54	Self-produced-1	255.12 ± 24.15	8.77 ± 3.51	1.45 ± 0.32	1.88±0.73	0.212±0.125	0.35±0.18
55	Self-produced-2	268.78 ± 37.49	12.00 ± 3.67	3.48 ± 0.30	1.78 ± 0.09	0.039 ± 0.006	0.52 ± 0.17
56	Self-produced-3	250.13 ± 26.04	3.87 ± 0.24	1.15 ± 0.14	1.74 ± 0.03	0.155 ± 0.003	0.15 ± 0.01
57	Self-produced-4	402.10 ± 38.38	55.49 ± 27.03	2.97 ± 0.16	10.32 ± 0.36	0.239 ± 0.045	1.52 ± 0.20
58	Self-produced-5	175.49 ± 16.92	6.61 ± 0.26	2.44 ± 0.16	1.69 ± 0.46	0.175 ± 0.031	0.25 ± 0.10

[a] Means ± SD ($n = 3$).

Table 1. Saccharification power and enzyme activities of Korean traditional *Nuruk* [21].

In Lee et al.'s study [21], 58 different kinds of traditional *Nuruk* were prepared, including 46 types of restored *Nuruk* mentioned in ancient documents. The saccharification power and glucoamylase, α-amylase, β-amylase, protease, and β-glucanase activities of each *Nuruk*s were reported. Among the 46 different kinds of restored and 12 types of collected *Nuruk*, the saccharification power values were the highest in *Bungok* (791 sp), followed by *Gokja* (421 sp), *Hwangguk* (226 sp), and *Backguk* (195 sp). The saccharification power of 12 kinds of commercial and self-produced *Nuruk* and their measured enzymeatic activities have been reported through Lee's study [21] that can be shown in **Table 1**. The range of saccharification power of the restored *Nuruk* was 85–565 sp. *Nuruk* with the highest saccharification power was shown in *Bungok* (565.5 sp), *Naebubijeon* (565.2 sp), and *Migok* (547.9 sp). This indicate that some restored *Nuruk* has a significantly higher saccharification power value than *Jinjugokja* (Keumkang wheat, 460.1 sp), which is a commercial *Nuruk*, and *ShinDaRi Nuruk* (477.2 sp), which is a self-produced *Nuruk*. The higher α-amylase activities of restored *Nuruk* were recorded in *Bungok* (65.53 U/g) and *Byunggok* (53.2 U/g), which was higher than *JinjuGokja* (26–38 U/g) showing a high activity among collected *Nuruk*. *Bungok* had higher α-amylase activity than self-manufactured *Nuruk*, *Igasubul* (55.49 U/g). Also, the β-amylase activity was the highest in *Jinjugokja* which is commercially available, and protease activity was highest in the self-manufactured *Igasubul* (10.32 μg/mL), followed by restored *Omegigok* (9.9 μg/mL). The α-amylase and β-amylase activities correlated significantly with the saccharification power value ($p < 0.001$). The correlation between the glucoamylase activity and saccharification power was also confirmed ($p < 0.05$). On the other hand, the activities of β-glucanase and protease in traditional *Nuruk* were not correlated with saccharification power value.

Author details

Jang-Eun Lee* and Jae-Ho Kim

*Address all correspondence to: jelee@kfri.re.kr

Traditional Alcoholic Beverages Research Team, Korea Food Research Institute, Gyeonggi-do, Republic of Korea

References

[1] Yu TS, et al. Bibliographical study on microorganisms of Nuruk (until 1945). Journal of the Korean Society of Food and Nutrition. 1996;**25**(1):170-179

[2] Back SY, et al. Assessment of the quality characteristics of mixed-grain Nuruk made with different fungal strains. The East Asian Society of Dietary Life. 2012;**22**(1):103-108

[3] Kim M, et al. Characteristics of Byeo-Nuruk according to the mixing ratio of wheat and the addition rate of moisture. The East Asian Society of Dietary Life. 2011;**21**(6):897-904

[4] Lee Y, So M. Effects of culture conditions of rhizopus sp. ZB9 on the production of saccharifying amylase during the preparation of rice Koji. The Korean Journal of Food and Nutrition. 2009;**22**(4):644-649

[5] Hong Y, et al. Microflora and physicochemical characteristics of Nuruk and main mashes during fermentation of a traditional andong soju. Foods and Biotechnology. 1997;**6**(4):297-303

[6] Kim H, et al. Characteristics of useful fungi isolated from traditional Korean Nuruk. Journal of the Korean Society of Food Science and Nutrition. 1997;**26**(5):767-774

[7] Kwon S, Shon J. Analysis of microbial diversity in Nuruk using PCR-DGGE. Korean Journal of Life Science. 2012;**22**(1):110-116

[8] Song SH, et al. Analysis of microflora profile in Korean traditional Nuruk. Journal of Microbiology and Biotechnology. 2013;**23**(1):40-46

[9] Yang S, et al. Aspergillus oryzae strains isolated from traditional Korean Nuruk: Fermentation properties and influence on rice wine quality. Food Science and Biotechnology. 2013;**22**(2):425-432

[10] Lee G, et al. Quality characteristics of Takju, Yakju, spirit made by pulse crop Nuruks. The Korean Journal of Culinary Research. 2015;**21**(3):232-247

[11] Lee J, Kang S, Cheong C. Quality characteristics of distilled alcohols prepared with different fermenting agents. Journal of the Korean Society for Applied Biological Chemistry. 2015;**58**(2):275-283

[12] Park J, Jeong J. Characteristics of takju (a cloudy Korean rice wine) prepared with nuruk (a traditional Korean rice wine fermentation starter), and identification of lactic acid bacteria in Nuruk. Korean Journal of Food Science and Technology. 2014;**46**(2):153-164

[13] Takamine K, et al. Characterization of Yeast for Soju (distilled spirits) from Korean traditional Nuruk. The Korean Journal of Mycology. 2015;**43**(3):196-199

[14] Kim M, et al. Diversity, saccharification capacity, and toxigenicity analyses of fungal isolates in Nuruk. The Korean Journal of Microbiology. 2014;**42**(3):191-200

[15] Kim M, et al. In-vitro anti-thrombosis activity of R4-Nuruk made from rhizopus oryzae KSD-815. Korean Journal of Microbiology and Biotechnology. 2015;**43**(2):169-174

[16] Lee S, et al. Inhibitory effects of ethanol extracts from Nuruk on oxidative stress, melanogenesis, and photo-aging. Mycobiology. 2012;**40**(2):117-123

[17] Son J, et al. Anti-adipogenic, anti-inflammatory, and anti-proliferative activities of extracts from lees and Nuruk. Journal of Life Science. 2015;**25**(7):773-779

[18] Bal J, et al. Pyrosequencing reveals bacterial diversity in Korean traditional wheat-based Nuruk. Journal of Microbiology. 2015;**53**(12):812-819

[19] Bal J, et al. Mycoflora dynamics analysis of Korean traditional wheat-based Nuruk. Journal of Microbiology. 2014;**52**(12):1025-1029

[20] Ponnusamy K, Lee S, Lee C. Time-dependent correlation of the microbial community and the metabolomics of traditional barley Nuruk starter fermentation. Bioscience Biotechnology and Biochemistry. 2013;**77**(4):683-690

[21] Lee J, et al. Restoration of the traditional Korean Nuruk and the brewing characteristics analysis. Journal of Microbiology and Biotechnology. 2017. in press. 2017;**27**(5):896-908

[22] Ribes J, BD. Vanover-sams CL, zygomycetes in human disease. Clinical Microbiology Reviews. 2000;**13**:236-301

[23] Battaglia E, et al. Carbohydrate-active enzymes from the zygomycete fungus rhizopus oryzae: A highly specialized approach to carbohydrate degradation depicted at genome level. BMC Genomics. 2011;**12**(1):1-12

[24] Yang S et al. Fungi associated with the traditional starter cultures used for rice wine in Korea. Journal of the Korean Society for Applied Biological Chemistry. 2011;**54**:933-943

[25] Norihiro T, et al. Isolation of a cDNA encoding aspergillus oryzae taka-amylase A: Evidence for multiple related genes. Gene. 1989;**84**:319-327

[26] Chang P, Ehrich K, Hua S. Cladal relatedness among aspergillus oryzae isolates and aspergillus flavus S and L morphotype isolates. International Journal of Food Microbiology. 2006;**108**:172-177

[27] Hong S, et al. The propoortion of non-aflatoxigenic strains of the aspergillus flavus/oryzae complex from Meju by analyses of the aflatoxin biosynthetic genes. Journal of Microbiology. 2013;**51**:766-772

[28] Bhatnagar D, et al. Understanding the genetics of regulation of aflatoxin production and Aspergillus flavus development. Mycopathology. 2016;**162**:155-166

[29] Bhatnagar D, Ehrich K, Cleveland T. Molecular genetic analysis and regulation of aflatoxin biosynthesis. Applied Microbiology and Biotechnology. 2003;**61**:83-93

[30] Eun-Gyung M, et al. Aflatoxin biosynthesis cluster gene cypA is required for G aflatoxin formation. Applied and Environmental Microbiology. 2004;**70**:6518-6524

[31] Wei D, et al. Molecular characterization of atoxigenic aspergillus flavus isolates collected in China. The Journal of Microbiology. 2014;**52**:559-565

[32] Lee M, Lee S, Yoon T. The bibliographical study on Nuruk. The East Asian Society of Dietary Life. 1994;**4**:19-29

[33] Lee, SB. Studies on enzymic sources and method of effective addition in fermentation of Yack-Tack-Joo Korean liquors. The Korean Journal of Microbiology. 1967;**5**:43-54

Section 4

Application of Brewing Industry Waste

Chapter 7

Exploitation of Brewing Industry Wastes to Produce Functional Ingredients

Anca Corina Fărcaş, Sonia Ancuța Socaci,
Elena Mudura, Francisc Vasile Dulf, Dan C. Vodnar,
Maria Tofană and Liana Claudia Salanță

Additional information is available at the end of the chapter

http://dx.doi.org/10.5772/intechopen.69231

Abstract

Nowadays, the consumers' global demand for healthier diets is steadily increasing, and the development of novel functional ingredients has become a focus of the food industry. On the other hand, the accumulation of huge amounts of food wastes every year has led to environmental degradation and especially to significant loss of valuable material that could otherwise be exploited as new health-promoting ingredients, fuels and a great variety of additives. In this respect, the biggest challenge of the current scientific world is to convert the underutilised by-products generated by the food and beverage industries into more profitable and marketable added value products which would also contribute significantly to meet the nowadays society needs. This chapter gives an overview regarding the possibility of exploiting the brewing industry wastes as sources of bioactive compounds in order to produce functional ingredients and products with added value.

Keywords: brewing wastes, bioactive compounds, innovative functional ingredients, waste management, complete valorisation

1. Introduction

The research on the recovery of new functional ingredients from natural sources is one of the most important challenges in food science and technology [1, 2]. In recent years, it has been noticed a significant increase in the number of research providing scientific evidence to support the hypotheses that phytochemicals recovered from agro-industrial wastes can provide a range of health benefits to the consumers. This fact has impacted the food and pharmaceutical industries, among others. The phytochemical extracts can be used either for their biological properties as ingredients for nutraceutical preparations or functional foods or for their food-quality-related properties [3].

In the food industry, the traditional significance of "processing" is associated with transformation of the initial raw material into a safe, nutritious and high-quality food product. However, in a modern bio-based society, food processing should also provide viable alternative models that combine food production with valorisation of waste and by-product, minimisation of energy consumption and environmental protection [4].

The brewing industries produce millions of tons of residues, which represent a management issue from both ecological and economical point of view. The accumulation of huge amounts of this biomass every year leads to environmental degradation and especially to significant loss of valuable material that could otherwise be exploited as food, fuels and a great variety of additives. The valorisation of brewing by-products can be achieved through the extraction of high-value components such as proteins, polysaccharides, fibres, flavour compounds and phytochemicals, which can be reused as nutritionally and pharmacologically functional ingredients [5–9]. Nowadays, the advances in scientific research support the idea that diet may fulfil nutritional needs and at the same time exert a beneficial role in protecting the human body of some diseases. The idea of health-promoting foods is not new: Hippocrates wrote 2400 years ago, "Let food be thy medicine and medicine be thy food." Thus, the production and consumption of functional foods have gained much importance as they provide a health benefit beyond the basic nutritional functions [10, 11]. The innovative approach in developing a new generation of functional ingredient and foods is focused on finding unconventional sources of bioactive compounds and optimising the most appropriate recovery system.

In this chapter, we will address the exploitation of brewing-derived by-products as sources of bioactive compounds and functional ingredients.

2. Bioactive compounds recovered from brewing wastes and their potential applications as functional ingredients

2.1. General overview

The recovery and reuse of the brewing industry by-products to extract functional compounds and develop new innovative products are a research direction of great interest and actuality from the perspective of food—health relation as well as from the environment protection and waste management perspective.

During production, beer alternately goes through three chemical and biochemical reactions: mashing, boiling and fermentation. In the mashing stage, malt starch is converted to fermentable sugars (mainly maltose and maltotriose) and non-fermentable sugars (dextrins), and proteins are partially degraded to polypeptides and amino acids. This enzymatic conversion stage produces sweet liquid called wort and a residual solid fraction called spent grains. After filtration, the wort is transferred to the brewing kettle where it is boiled with the addition of hops. During this process, the bitter and aromatic hop components will confer typical beer qualities, such as bitter taste, flavour and foam stability. At the boiling end, the liquid extract is separated from the spent hops to be further processed. A fraction of the hop components end up in the trub (a precipitation product of the wort boiling process that may include insoluble hop materials, condensation products of hop polyphenols and wort proteins, and isomerised hop acids). During fermentation, the yeast cells will convert the fermentable sugars to ethanol and carbon dioxide. At the end of this stage, most of the cells are collected as spent yeast [7, 12, 13]. According to the technological process shown schematically in **Figure 1**, the main by-products generated in the brewing process are spent grain, spent hops and trub and spent yeast.

There is a wide range of extraction techniques used for the isolation and purification of the bioactive compounds from brewing wastes, some of them based on new emerging techniques. The extraction of the high-value components must be economically feasible to perform. This objective can be achieved by separating the components of interests through individual and/or combined physical and biochemical approaches in order to provide a range of components, all of which would contribute to achieving whole-waste exploitation [6, 9]. The extraction conditions are extremely important, due to their effects on the release of compounds from the matrix into the medium and also due to structural changes that may occur and alter the expected properties. Thus, the optimisation of the existing methodologies and development of new extraction methods to increase the extraction yield, the selectivity for a certain compound, to protect their functionality or to extend the applicability in the food industry are of utmost importance. Also, in order to increase the overall sustainability of the reuse

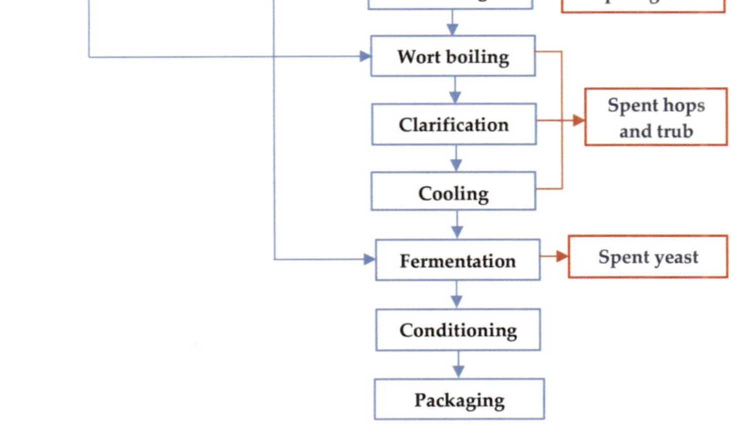

Figure 1. Schematic representation of the brewing process and points where the main by-products are generated.

of components of by-products in food life cycle, it will be necessary to apply and promote novel eco-friendly extraction technologies capable of reducing the solvent consumption and to ensure the environmental protection [6, 14, 15].

2.2. Brewers' spent grain

The brewers' spent grain (BSG) is the main solid waste produced in large quantities by the beer industry, resulted after mashing and filtration stage. This insoluble material basically consists of the barley grain husk in the greatest proportion, minor fractions of pericarp and fragments of endosperm and other residual compounds not converted into fermentable sugars by the mashing process [8, 16, 17].

2.2.1. General characterisation of BSG

In the brewery, the malted barley is milled and mixed with water in the mash tun, and the temperature of mash is slowly increased from 37 to 78°C in order to promote the enzymatic conversion of malt constituents. After the saccharification process, the clear sweet wort is separated from the solid components—the spent grain. The wort is then transferred to the wort kettle, whereas the spent grain is removed from the lauter tun [18].

As described in the literature, the chemical composition of BSG is variable according to the barley variety and harvest time, malting and mashing conditions, type and quality of secondary raw materials added in the brewing process. A major influence is that different barley cultivars are used as the malt source for lager and ale beer. In general, ale malt is kilned at a higher temperature, whereas lager malts are derived from barley with higher protein content [8, 12, 19, 20]. Recent research on the evaluation of the BSG biomass showed surprising results both in terms of the variety of classes of compounds and the quantity of the functional part. Even if BSG chemical composition is dependent on the intrinsic and extrinsic factors mentioned above, it contains appreciable amounts of valuable compounds (proteins, lipids, carbohydrates, polyphenols, minerals) that remain unexploited in the brewing process [5, 7, 8, 17].

BSG is an important by-product from the brewing process, representing up to 30% (w/w) of the starting malted grain. It is estimated that worldwide the annual output is around 30 million tons, about 200 tons of wet BSG (70–80% water content) being produced per 10.000 hl of beer. Traditionally, this material is sold as animal feed or discarded [8, 16, 21, 22]. Due to the significant amount produced annually, the low market value, environmental awareness and the recognition that BSG may represent a nutritionally valuable co-product, efforts should be increasingly focused on its valorisation [17, 23, 24].

The high initial water content of fresh BSG (75–80%) and the presence of considerable levels of polysaccharide, residual fermentable sugars and proteins make BSG very susceptible to microbial degradation within a few days [25]. Microbiological stabilisation is an imperative that should be designed into BSG processing systems in order to avoid the growth of microorganisms. The implications for microbiological spoilage by a resident microflora might affect also the potential to use BSG as a reliable food-grade industrial feedstock for value-added downstream processing [20]. One of the most common and economically feasible methods used for preservation is the drying of BSG, thus reducing the water content and lowering the microbiological activity. This preservation method is also interesting in terms of reducing the volume of the product and, therefore, decreasing transport and storage costs (**Figure 2**) [19].

2.2.2. Bioactive compounds and potential applications

2.2.2.1. Proteins and amino acids

The protein content of barley varies from 8 to 15%. During malting, barley proteins are partially degraded to amino acids and small peptides by the endogenous barley peptidases. However, most of malt proteins are not dissolved in mashing but 74–78% of protein remains insoluble in the spent grains. As a result, BSG has a high content (18–35.4%, w/w) of quality protein [16, 26–28]. For a by-product to be considered as a source of quality protein, it must contain a well-balanced essential amino acid composition [9, 29]. In BSG, the essential amino acids represent approximately 30% of the total protein content. Lysine, known to be the limiting amino acid in cereal derived foods, accounts for 14.3% of the total BSG protein content. Other amino acids in significant quantity are: leucine, phenylalanine, isoleucine, threonine and tryptophan [21]. Depending on the raw materials used in brewing process (unmalted grain, corn, barley, wheat, rice), protein composition can vary significantly thereby modifying the essential amino acids profile of BSG. Owing to its protein-rich composition, BSG has the potential to be utilised in a manner similar to whey protein, providing health benefits for consumers [7, 21, 30, 31]. Essential amino acids of BSG derived from 100% barley malt expressed as a percentage of total protein [21] are listed in **Table 1**.

The lack of solubility of BSG proteins presents a barrier for their more extensive use in food processes and products. Protein hydrolysates from agricultural crops have already demonstrated bioactive effects which support their potential use as functional food ingredients. In addition, these products can have numerous properties indispensable for the food industry such as emulsifying agents, film forming properties, flavour binding, viscosity increase by

Figure 2. Fresh (left) and dried (right) BSG.

Essential amino acids (% of total protein)	Barley	Malt	BSG
Lysine	2.52	3.69	14.31
Leucine	0.30	0.29	6.12
Phenylalanine	0.20	0.21	4.64
Isoleucine	0.17	0.17	3.31
Threonine	0.01	0.02	0.71
Tryptophan	0.01	nd	0.14

Table 1. Essential amino acids content of BSG derived from 100% barley malt.

binding the water and gelation properties. To expand the potential applications of insoluble proteins, chemical and enzymatic hydrolysis can be applied [16, 27, 32]. However, BSG protein fraction can be a valuable substrate for enzymatic hydrolysis to produce hydrolysates with biological properties. BSG protein hydrolysates are also of high importance when considering incorporation into food products, particularly with respect to their techno-functional properties, of which solubility, emulsifying properties, immune-modulatory effects and antimicrobial activity are very important. Protein hydrolysis changes the molecular weight, charge and exposure of hydrophobic groups and amino acid side chains, which alters solubility, viscosity, sensory properties and emulsifying and foaming behaviour [17, 21, 27, 33–35].

In the present, studies have already demonstrated that food with incorporated protein hydrolysates derived from BSG possesses anti-inflammatory activity [32, 36]. Moreover, the incorporation of chitosan into the brewers' spent grain protein had as result a composite film with antimicrobial and antioxidant activities which can be used in packaging materials [34].

2.2.2.2. *Polysaccharides*

The main constituents of BSG include fibres and proteins, which are staple nutritional components in the human diet and thus make this material very attractive for improving the nutritional value of foods. In addition, several components that are constituents of BSG, such as arabinoxylans, β-glucans and phenolic compounds (e.g. hydroxycinnamic acid), have gained increasing attention due to their potential health benefits. The determination of the beneficial implications of dietary fibre in a wide variety of food products has resulted in fibre being regarded as a "functional" ingredient. The physiological responses to fibre consumption have been well documented in relation to reduced risks of chronic diseases [37]. BSG contains the husk and the outer layers of the barley kernel thus being a heterogeneous material rich in arabinoxylan (22 – 28%), cellulose (17 – 25%) and lignin (12–28%). This lignocellulosic material is constituted by several polysaccharides, which can be degraded into their corresponding constituents by hydrolytic procedures (hydrothermal, enzymatic or acidic). On hydrolysis, cellulose yields glucose, whereas the hemicellulose yields xylose, arabinose, mannose, galactose and the acids, acetic and hydroxycinnamic (ferulic and p-coumaric) [13, 22, 38]. The released monosaccharides can be further subjected to a fermentation process to generate valuable products (e.g. xylitol, a healthy sweetener used in food industry) [39]. Also, the arabinoxylans are considered dietary fibres with a broad range of potential uses as functional ingredients in food products. In addition, for maximising the valorisation efficiency of BSG, an innovative integrated process that sequentially extracts the proteins and arabinoxylans was recently developed [16].

An alternative of reuse of dried BSG is as flour incorporated in food products [19, 28]. Fibre is suitable in meat products because it retains water, decreases cooking losses and has a neutral flavour. Therefore, having a high potential as a source of dietary fibre BSG can be used as a fat substitute for producing high-dietary fibre and low-fat meat products, reduces the number of synthetic antioxidants needed to be added and increases the health-promoting properties of the frankfurters [37]. Also, BSG was successfully used in smoked sausages meat products to partially replace the animal protein [40]. Processing BSG into flour represents a viable alternative for its use as a functional ingredient in bakery products. Different percent of BSG flour added to bakery products resulted in increase of total dietary fibre, total protein content, lipids, minerals and water holding capacity of the final product. The substitution of wheat flour with 5–20% BSG resulted in bread prototypes with enhance nutritional value and with pleasant flavour characteristics imparted by the specific volatile compounds [19, 21, 28, 41–43].

Recently, BSG was investigated as potential raw material for cellulose nanofibre production. Anticipated applications of nanocellulose range from food (e.g. emulsion/dispersion) to medical, cosmetic, pharmaceutical, hygiene/absorbent products and even usage in various nanocomposites and paper applications [44–46].

Due to the composition rich in sugars and nutritional factors, hydrolysates produced from BSG can be used in fermentative processes to produce several compounds of industrial interest. The BSG biomass can also be exploited by microbial-processing in order to obtain valuable enzymes and organic acids. These valuable chemicals can be further exploited as raw materials for other processes or as functional ingredients for development of new generations of added-value food products and not only. Some examples include the use of the sugar rich hydrolysate as fermentation medium for the production of ethanol by *Saccharomyces cerevisiae*, xylitol by *Candida guilliermondii*, xylitol, arabitol, ethanol and glycerol by *Debaryomyces hansenii*, and lactic acid by *Lactobacillus delbrueckii*, *Lactobacillus pentosus* or *Lactobacillus rhamnosus* [13, 47, 48].

2.2.2.3. *Phenolic compounds*

The recent worldwide tendency to avoid or at least decrease the use of synthetic additives has created a need for alternative cheap, renewable, natural and possibly safer sources of natural compounds with antioxidant and antimicrobial activities to stabilise foods against oxidative rancidity and microbial spoilage [49]. Having strong antioxidant activity, when ingested, the antioxidants protect the human body from the damaging actions of the reactive oxygen species and thus lowering the risk of several chronic diseases (cardiovascular diseases, diabetes, cancer) [7, 9, 33, 50, 51].

Beer contains a large variety of phenolic components derived from the biotechnological fermentation of barley malt (70%) and hops (30%) that are responsible for the overall antioxidant activity of the beverage. These compounds play an important role in flavour, colour and shelf life of beer [52]. Since most of the phenolic compounds of the barley grain are contained in the husk and hydroxycinnamic acids accumulate in the cell walls, BSG is a potentially valuable and inexpensive source of phenolic acids. Ferulic acid (1860–1948 mg/g) and p-coumaric (565–794 mg g^{-1}) are the most abundant phenolic acids in BSG, followed by sinapic, caffeic and syringic acids [7, 9, 17, 33]. The content in polyphenols is influenced not only by the extraction technique but also by factors such as barley cultivar, malting conditions and the presence or absence of the hull. Depending on the solvent used for extraction, studies reported for total phenols in BSG values ranging from 2.14 to 9.90 mg GAE/g and a flavonoid content varying between 0.02 and 4.61 mg QE/g [50]. **Table 2** overviews the results of the authors' own research regarding the total phenols, flavonoids and radical scavenging activity values for barley, malt and BSG samples.

However, the application of BSG to food is still limited, since it can impart unpleasant flavours and aromas. This negative effect was associated with high quantities of compounds such as 2-butyl-1-octanol, 3-methyl-butanal, 2-heptane, butanal, benzene and 2, 3-butanedione, responsible for its characteristic unpleasant odour [53]. In a recent study, the polyphenols and flavonoids compounds extracted by supercritical CO_2 from BSG were microencapsulated to mask their unpleasant and bitter taste and simultaneously to preserve the stability of polyphenols or other bioactive compounds. The tests on fish-burger formulation with microencapsulated polyphenols showed to have about 30% of phenolic and about 50% of flavonoid content more than the control sample and a better antioxidant activity [53].

2.2.2.4. Lipids and fatty acids

Another important BSG macro-nutrient includes lipids and fatty acids. The high amounts of lipids in BSG make this material an interesting feedstock for the production of high value-added lipids in the context of the so-called lignocellulose biorefinery [5]. The lipids in barley are located in the endosperm and embryo, as their role is to provide nutrients and energy for the new, germinating barley plant. Although the endosperm is almost completely solubilised in mashing, most of the lipids remain with spent grains and are not transferred to wort [54–56]. According to recent studies [5, 8, 55, 57], the total lipid contents (TLs) of BSG

Sample	Total phenols (mg GAE/100 g fw)	Flavonoids (mg QE/100 g fw)	DPPH inhibition (%)
Barley	133.93	6.17	43.17
Pilsner malt	148.42	5.28	46.36
Caramunich malt	256.42	10.72	57.87
Carafa malt	335.88	8.97	42.07
Dried BSG	284.20	13.16	55.95
Lyophilised BSG	291.47	10.35	53.78

Table 2. Total phenols, flavonoids and radical scavenging activity values for barley, malt and BSG.

varied between 5.40 and 11.00% (dry-mass). Only minor changes in fatty acid composition occur during malting and mashing, and therefore, the fatty acid composition of BSG is similar to that of barley [8, 54].The predominant lipid classes identified in the total lipids of BSG were triacylglycerols (TAG) (55–67% total lipids), followed by free fatty acids (FFA) (18–30%), diacylglycerols (DAG) (7.7–5.7%), monoacylglycerols (MAG) (1.7%), phospholipids (PL) (9.1%) and steroid compounds (SC) (hydrocarbons, ketones, free sterols, sterol esters and sterol glycosides) (5%) [5, 55]. The high amount of free fatty acids in BSG can be attributed to the endogenous lipase that is able to release the free fatty acids from triacylglycerols and polar lipids during malting and mashing [5, 55]. **Table 3** contains the results of the authors' own research regarding the fatty acids profile of barley, malt and BSG [8].

Regarding the fatty acids composition (GC-MS analysis) of lipids from BSG, the studies showed that the most abundant was linoleic acid (18:2, *n*–6) that accounted for 50–51.50% of all identified fatty acids, followed by palmitic (16:0) (25–26% of total fatty acids) and oleic acids (18:1, *n*–9) (12–13% of total fatty acids). Small amounts of other fatty acids, such as stearic (18:0) and linolenic (18:3, *n*–3) acids, were also reported [8, 55]. The elevated level of linoleic acid (18:2, *n*–6) from the BSG lipids is comparable to those of "linoleic acid-rich" vegetable oils, such as grape seed, hemp seed and wheat germ oils [58]. This acid is an *n*–6 essential fatty acid, which can be used in pharmaceutical and cosmetic products, and is considered to influence the metabolic processes in the skin and to promote the activity of different lipophilic vitamins, such as A and E [5].

2.3. Brewers' spent yeast

The brewers' spent yeast (BSY) is another brewing by-product that merits considerable attention, due to the large quantity produced (is the second largest by-product from breweries) and its rich chemical composition.

Fatty acids (% of total fatty acids)	Barley	Pilsner malt	Caramunich malt	Carafa malt	Dried BSG
∑SFAs	24.57	25.87	23.68	25.90	29.78
∑MUFAs	17.36	14.26	17.25	18.23	14.53
∑PUFAs	58.07	59.86	59.07	55.87	55.69
∑*n*–3 PUFAs	4.82	5.11	5.78	4.37	5.18
∑*n*–6 PUFAs	53.25	54.75	53.29	51.50	50.51
n–6/*n*–3	11.04	10.71	9.22	11.78	9.75
PUFAs/SFAs	2.36	2.31	2.49	2.16	1.87
∑VLCSFA (≥20 C)	0.89	1.01	0.63	0.62	0.96
Total lipids (% of dw)	2.96	2.55	2.31	2.74	6.61

SFAs, saturated fatty acids; MUFAs, monounsaturated fatty acids; PUFAs, polyunsaturated fatty acids and VLCSFAs, very long-chain saturated fatty acids.

Table 3. Total lipids composition and the major lipid fractions from barley, malt and BSG samples analysed by GC-MS.

2.3.1. General characterisation of BSY

Brewing yeast *Saccharomyces cerevisiae* is the technological biocatalyst, which produces beer from fermentable substrate by alcoholic fermentation. After 10–15 successive fermentation batches, the yeast, due to increasing contamination, loses its viability and vitality and is no longer proper for making beer [59]. BSY is the second major by-product of the brewing industry with environmental impact due to the disposal of a large quantity of biomass (1 hl of beer generates 2.0–4.0 kg of BSY) [60, 61]. BSYS can be collected from fermentation and storage tanks, the yeast storage plant and from the filter line. The quantity and quality of biomass harvest at the end of brewing production depend on the pitching rate, the yeast viability, the yeast strain, the purity of the yeast culture, the wort composition, the particulate content of the wort, the timing and extent of wort aeration/oxygenation, the fermentation conditions and the plant capability [62]. Management of BSY is one of the most important concerns of breweries. BSY contains liquids in large quantities (85–90%), which makes handling and disposal difficult and expensive. As a good practice, the brewers concentrate the waste yeast (to 22–25% dry matter) and also recover the beer to reduce losses [25, 63, 64].

2.3.2. Bioactive compounds and potential applications of BSY

The major chemical compounds of BSY are represented by carbohydrates, proteins, free amino acids, ash, vitamins and fatty acids. The predominant amino acids found in brewer's spent yeast proteins are leucine, lysine, tyrosine, arginine, cysteine, histidine, isoleucine, methionine, phenylalanine, threonine, tryptophan and valine. Thus, BSY is an excellent source of high-quality protein, comparable in value with soy protein. Also, reported high values of glutamic acid and glutamine contents, increase the potential use of BSY extract in food industry as a "hidden ingredient" of natural monosodium glutamate, which is known to provide the typical"umami" aroma, very similar to meat aroma [25, 65, 66]. The inner layer of the BSY cell wall contains β-glucans 8% (w/w dry weight), and the external layer is formed by mannoproteins [67]. These two classes of compounds have immunomodulatory, antimutagenic and anticarcinogenic activities, being also utilised in cosmetic products, and lately in food industry as natural emulsifiers [68–70]. The physicochemical properties of β-glucans depend on the characteristics of their primary structure, including linkage type, degree of branching, molecular weight and conformation. β-glucans from yeast, which consist of a (1, 3)-β-linked backbone with small numbers of (1, 6)-β-linked side chains, are essentially known for their immune-modulating effects [71, 72]. They also stimulate the skin cell response to combat free radicals thus significantly delaying ageing process. The European Food Safety Authority (EFSA) has already approved the use of *Saccharomyces* β-glucans as a new food ingredient and suggests a use ranging between 50 and 200 mg per serving [73]. β-glucans from BSY have potential applications in food industry as food thickeners, fat replacer, dietary fibres, viscosity imparting agents, emulsifiers and films [74]. For example, due to its lower calories content, BSY can be a valuable source of cheap easily assailable fibre, with recently proved prebiotic effect, which makes it of interest for pastry industry, in order to develop value added products [66, 75].

Nevertheless, BSY has a strong antioxidant activity, comparable with that of teas, due to phenolic compounds adsorbed from malt and hop in the brewing process. BSY contain a high level of phenolic compounds in both the free and bounded forms: gallic acid, protocatechuic acid, (±) catechin, p-coumaric, ferulic and cinnamic acids, which makes BSY a potential functional ingredient [76, 77].

Other compounds derived from hops include α- and β- acids, which have a strong antimicrobial activity [78]. The α-acid content ranges between 167 and 2074 μg/g, with an average related to the total hop acid between 487 and 2557 μg/g. When centrifugation is used for yeast separation, the amount of hop acids is higher, demonstrating the BSY affinity for these compounds [79].

The lipid fraction of the BSY accounted for 4.4% of dry biomass, 58% of which were neutral lipids. Mono-, di- and triacylglycerols, squalene, lanosterol, ergosterol, steryl esters and free fatty acids were identified in the neutral lipid fraction. Although brewers' yeast does not belong to the so-called lipid yeasts, the high content of squalene gives reason for additional exploitation of this by-product [80].

BSY consumption as protein source for human nutrition is limited by the high level of nucleic acids (6–15%) which can cause increase in the level of uric acid in the blood and tissues. This restricts the use of BSY to the status of a dietary supplement as powders, flakes, tablets or in liquid form, rich in bioactive compounds: vitamins, especially B vitamins and minerals (calcium, phosphorus, potassium, magnesium, copper, iron, zinc, manganese, selenium and chromium). In order to be used as dietary supplement, BSY has to be subjected to a debittering process that can conventionally be achieved by washing with an alkaline aqueous solution or by water vapour distillation with or without an organic solvent treatment [81, 82].

Autolysed BSY extract occurs by the natural action of endogenous enzymes when cells complete their growth cycle. The cell wall gets disrupted as the yeast's enzymes break down proteins, releasing amino acids, salts and carbohydrates. The soluble portions are separated from the insoluble components by centrifugation and several filtration steps, including ultrafiltration. The final product is either stored in liquid or paste form or may be spray-dried to a powder [25, 63]. BSY hydrolysates are obtained by acidic or enzymatic (proteolytic enzymes) hydrolysis. The BSY extracts manufactured by autolysis and hydrolysis are used as a functional ingredient in a variety of processed foods: meat paste, soups, sauces, snacks and vegetarian foods, but due to the high salt content, it may have limited uses [83].

Yeast extract from BSY could be used in a large variety of food as flavour enhancers [63, 84]. The intracellular enzymes nucleases produce nucleotides and nucleosides, of which 5′-guanosine monophosphate and 5′-inosine monophosphate act as flavour enhancing—the so-called umami effect proteases breakdown the proteins into smaller polypeptides and sulphur amino acids that provide komumi taste, continuity, mouthfulness and thick flavour. Several sulphur-containing compounds, identified as S-allyl-cysteine sulphoxide (alliin) and glutathione (GSH, γ-Glu-Cys-Gly), were responsible for this effect [85].

2.4. Brewers' spent hops and trub

Only 15% of the hops constituents will be retrieved in the beer, whereas 85% will become spent hop material. A fraction of the hop components will end up in the trub, mainly when hop powdered pellets or extracts are used in the brewing process. The hot trub is a precipitation product of the wort boiling process that includes: insoluble hop materials, condensation products of hop polyphenols and wort proteins and isomerised hop acids adsorbed on the trub solids [12, 13]. Compared to BSG, the direct use of spent hops as feed supplement is not desirable due to the presence of 2-methyl-3-buten-2-ol, which is the product of bitter acid degradation and has hypnotic-sedative properties. Traditionally, spent hops have been used as a fertiliser and soil conditioner, due to the high nitrogen content or mixed with spent grain and sent to animal feeding. However, there are several compounds can be recovered from spent hops, such as flavours, saccharides and organic acids, which can be obtained after oxidation or hydrolysis of this material. Among these compounds, the hop acids, particularly, have antibacterial potential being a safe alternative to control bacteria in ethanol fermentations and able to efficiently replace antibiotics in ethanol production [12, 13, 86].

3. Concluding remarks and future trends

The idea of converting the brewing waste into functional ingredients is an area of research with huge potential and opportunities. Recent advances in biotechnology ensure that brewing industry by-products are no longer regarded as a waste but rather a feedstock for producing a new generation of added-value products. Based on this, it is an undeniable fact that brewing residues have their own potential for sustainable reuse through biotechnological processes. The recent findings highlighted the potential reuse of brewery by-product and led to the idea that multidisciplinary approaches should be implemented in order to develop integrated biorefineries. Despite the continuous progress in the recent years in this area, there are still the need and a priority to develop/adapt modern and efficient methods for extraction of these bioactive compounds.

As economic impact, using the BSG by-product, which has a low monetary value, as a high-nutrient biomass, will enhance the economic potential of breweries and improve the dietary attributes of food formulations. The recovered bioactive compounds and functional ingredients are also of great interest for food, pharmaceutical industry (e.g. antimicrobial activity, carrier agents, controlled release, immune-modulatory effects), cosmetics, agriculture, chemical industry and not only. The social impact of these actions refers to the fact that the complex recovery of bioactive compounds and new functional ingredients is aimed to be an efficient and at the same time affordable alternative, for all social categories, to complete their diet with an appreciable number of nutrients.

Taking into consideration all the above, the future trend in exploitation dewaste generated in the beer making process is represented by the development and optimisation of different integrated extraction system of biologically active compounds, thus maximising the valorisation efficiency of brewing waste, insufficiently exploited until now.

Acknowledgements

This work was supported by a grant of the Romanian National Authority for Scientific Research and Innovation, CNCS–UEFISCDI, project number PN-II-RU-TE-2014-4-0842.

Author details

Anca Corina Fărcaş[1], Sonia Ancuța Socaci[1*], Elena Mudura[1], Francisc Vasile Dulf[2], Dan C. Vodnar[1], Maria Tofană[1] and Liana Claudia Salanță[1]

*Address all correspondence to: sonia.socaci@usamvcluj.ro

1 Faculty of Food Science and Technology, University of Agricultural Sciences and Veterinary Medicine Cluj-Napoca, Romania

2 Faculty of Agriculture, University of Agricultural Sciences and Veterinary Medicine Cluj-Napoca, Romania

References

[1] Femenia A. High-value co-products from plant foods: Cosmetics and pharmaceuticals. In: Waldron K, editors. Handbook of Waste Management and Co-product Recovery in Food Processing. Cambridge, England: Woodhead Publishing Limited;. 2007, pp. 470-501

[2] Helkar PH, Sahoo AK, Patil NJ. Review: Food industry by-products used as a functional food ingredients. International Journal of Waste Resources. 2016;**6**(3):248-253. DOI: 10.4172/2252-5211.1000248

[3] Barberan FAT. High-value co-products from plant foods: Nutraceuticals, micronutrients and functional ingredients. In: Waldron K, editors. Handbook of Waste Management and Co-product Recovery in Food Processing. Cambridge, England: Woodhead Publishing Limited; 2007, pp. 449-469

[4] Lin CS, Koutinas AA, Stamatelatou K, Mubofu EB, Matharu AS, Kopsahelis N, Pfaltzgraff LA, Clark JH, Papanikolaou S, Kwan TH, Luque R. Current and future trends in food waste valorization for the production of chemicals, materials and fuels: A global perspective. Biofuels, Bioproducts and Biorefinery. 2014;**8**(5):686-715. DOI: 10.1002/bbb.1506

[5] Del Rio JC, Prinsen P, Gutiérrez A. Chemical composition of lipids in brewer's spent grain: A promising source of valuable phytochemicals. Journal of Cereal Science. 2013;**58**(2):248-254. DOI: 10.1016/j.jcs.2013.07.001

[6] Baiano A. Recovery of biomolecules from food wastes—A review. Molecules. 2014;**19**(9): 14821-14842. DOI: 10.3390/molecules190914821

[7] Mussatto SI. Brewer's spent grain: A valuable feedstock for industrial applications. Journal of the Science of Food and Agriculture. 2014;**94**(7):1264-1275. DOI: 10.1002/jsfa.6486

[8] Farcas AC, Socaci SA, Tofana M, Dulf FV, Mudura E, Diaconeasa Z. Volatile profile, fatty acids composition and total phenolic content of brewers' spent grain by-product with potential use in the development of new functional foods. Journal of Cereal Science. 2015;**64**:34-42. DOI: 10.1016/j.jcs.2015.04.003

[9] Socaci SA, Farcas AC, Vodnar DC, Tofana M. Food wastes as valuable sources of bioactive molecules. In: Naofumi Shiomi, editors. Superfood and Functional Food—The Development of Superfoods and Their Roles as Medicine. Rijeka, Croatia: InTech; 2017, pp. 75-93. DOI: 10.5772/66115

[10] Otles S, Cagindi O. Safety considerations of nutraceuticals and functional foods. In: McElhatton A, Sobral PJA, editors. Novel Technologies in Food Science. New York, US: Springer; 2012, pp. 121-136

[11] Corbo MR, Bevilacqua A, Petruzzi L, Casanova FP, Sinigaglia M. Functional beverages: The emerging side of functional foods commercial trends, research, and health implications. Comprehensive Reviews in Food Science and Food Safety. 2014;**13**(6):1192-1206. DOI: 10.1111/1541-4337.12109

[12] Huige NJ. Brewery by-products and effluents. In: Priest FG, Stewart GG, editors. Handbook of Brewing, 2nd ed. London, UK: CRC Press; 2006. DOI: 10.1201/9781420015171.ch18

[13] Mussatto SI. Biotechnological potential of brewing industry by-products. In: Nigam PS, Pandey A, editors. Biotechnology for Agro-industrial Residues Utilisation. New York, US: Springer; 2009, pp. 313-326. DOI: 10.1007/978-1-4020-9942-7_16

[14] Escarpa A, Gonzalez MC. An overview of analytical chemistry of phenolic compounds in foods. Critical Reviews in Analytical Chemistry. 2001;**31**(2):57-139. DOI: 10.1080/20014091076695

[15] Azabou S, Abid Y, Sebii H, Felfoul I, Gargouri A, Attia H. Potential of the solid-state fermentation of tomato by products by Fusarium solani *pisi* for enzymatic extraction of lycopene. LWT-Food Science and Technology. 2016;**68**:280-287. DOI: 10.1016/j.lwt.2015.11.064

[16] Vieira E, Rocha MAM, Coelho E, Pinho O, Saraiva JA, Ferreira I, Coimbra MA. Valuation of brewer's spent grain using a fully recyclable integrated process for extraction of proteins and arabinoxylans. Industrial Crops and Products. 2014;**52**:136-143. DOI: 10.1016/j.indcrop.2013.10.012

[17] Lynch KM, Steffen EJ, Arendt EK. Brewers' spent grain: A review with an emphasis on food and health. Journal of Institute of Brewing and Distilling. 2016;**122**(4):553-568. DOI: 10.1002/jib.363

[18] Linko M, Haikara A, Ritala A, Penttila M. Recent advances in the malting and brewing industry. Journal of Biotechnology. 1998;**65**(2-3):85-98. DOI: 10.1016/S0168-1656(98)00135-7

[19] Santos M, Jiménez JJ, Bartolomé B, Gómez-Cordovés C, del Nozal MJ. Variability of brewer's spent grain within a brewery. Food Chemistry. 2003;**80**(1):17-21. DOI: 10.1016/S0308-8146(02)00229-7

[20] Robertson JA, I'Anson KJA, Treimo J, Faulds CB, Brocklehurst TF, Eijsink VGH, Waldron KW. Profiling brewers' spent grain for composition and microbial ecology at the site of production. LWT-Food Science and Technology. 2010;**43**(6):890-896. DOI: 10.1016/j.lwt.2010.01.019

[21] Waters DM, Jacob F, Titze J, Arendt KE, Zannini E. Fibre, protein and mineral fortification of wheat bread through milled and fermented brewer's spent grain enrichment. European Food Research and Technology. 2012;**235**(5):767-778. DOI: 10.1007/s00217-012-1805-95

[22] Niemi P, Faulds CB, Sibakov J, Holopainen U, Poutanen K, Buchert J. Effect of a milling pre-treatment on the enzymatic hydrolysis of carbohydrates in brewer's spent grain. Bioresource Technology. 2012;116:155-160. DOI: 10.1016/j.biortech.2012.04.043

[23] Buffington J. The economic potential of brewer's spent grain (BSG) as a biomass feedstock. Advances in Chemical Engineering and Science. 2014;**4**(3):308-318. DOI: 10.4236/aces.2014.43034

[24] Ravindran R, Jaiswal AK. Exploitation of food industry waste for high-value products. Trends in Biotechnology. 2016;**34**(1):58-69. DOI: 10.1016/j.tibtech.2015.10.008

[25] Mathias TR, de Mello PM, Sérvulo EF. Solid wastes in brewing process: A review. Journal of Brewing and Distilling. 2014;5(1):1-9. DOI: 10.5897/JBD2014.0043

[26] Jones BL, Budde AD. How various malt endoproteinase classes affect wort soluble protein levels. Journal of Cereal Sciences. 2005;**41**(1):95-106. DOI: 10.1016/j.jcs.2004.09.007

[27] Celus I, Brijs K, Delcour JA. Enzymatic hydrolysis of brewers' spent proteins and technofunctional properties of the resulting hydrolysates. Journal of Agricultural and Food Chemistry. 2007;**55**(21):8703-8710. DOI: 10.1021/jf071793c

[28] Farcas A, Socaci S, Tofana M, Muresan C, Mudura E, Salanta L, Scrob S. Nutritional properties and volatile profile of brewer's spent grain supplemented bread. Romanian Biotechnological Letters. 2014;**19**(5):9705-9714

[29] Oreopoulou V, Tzia C. Utilization of plant by-products for the recovery of proteins, dietary fibers, antioxidants, and colorants. In: Oreopoulou V, Russ W, editors. Volume 3. Utilization of By-products and Treatment of Waste in the Food Industry. New York, US: Springer; 2007, pp. 209-232. DOI: 10.1007/978-0-387-35766-9_11

[30] Reis SF, Coelho E, Coimbra MA, Abu-Ghannam N. Influence of grain particle sizes on the structure of arabinoxylans from brewer's spent grain. Carbohydrate Polymers. 2015;**130**:222-226. DOI: 10.1016/j.carbpol.2015.05.031

[31] Severini C, Azzollini D, Jouppila K, Jussi L, Derossi A, De Pilli T. Effect of enzymatic and technological treatments on solubilisation of arabinoxylans from brewer's spent grain. Journal of Cereal Science. 2015;**65**:162-166. DOI: 10.1016/j.jcs.2015.07.006

[32] Crowley D, O'Callaghan Y, McCarthy A, Connolly A, Piggott CO, FitzGerald RJ, O'Brien NM. Immunomodulatory potential of a brewers' spent grain protein hydrolysate incorporated into low-fat milk following in vitro gastrointestinal digestion. International Journal of Food Science and Nutrition. 2015;**66**(6):672-676. DOI: 10.3109/09637486.2015.1077788

[33] McCarthy AL, O'Callaghan YC, Connolly A, Piggott CO, FitzGerald RJ, O'Brien NM. In vitro antioxidant and anti-inflammatory effects of brewers' spent grain protein rich isolate and its associated hydrolysates. Food Research International. 2013;**50**(1):205-212. DOI: 10.1016/j.foodres.2012.10.022

[34] Lee J, Lee J, Yang HJ, Song KB. Preparation and characterization of brewer's spent grain protein-chitosan composite films. Journal of Food Science and Technology. 2015;**52**(11):7549-7555. DOI: 10.1007/s13197-015-1941-x

[35] Vieira E, Teixeira J, Ferreira IMPLVO. Valorization of brewers' spent grain and spent yeast through protein hydrolysates with antioxidant properties. European Food Research and Technology. 2016;**242**(11):1975-1984. DOI: 10.1007/s00217-016-2696-y

[36] McCarthy AL, O'Callaghan YC, Connolly A, Piggott CO, FitzGerald RJ, O'Brien NM. Brewers' spent grain (BSG) protein hydrolysates decrease hydrogen peroxide (H_2O_2)-induced oxidative stress and concanavalin-A (con-A) stimulated IFN-gamma production in cell culture. Food and Function. 2013;**4**(11):1709-1716. DOI: 10.1039/c3fo60191a

[37] Ozvural EB, Vural H, Gokbulut I, Ozboy-Ozbas O. Utilization of brewer's spent grain in the production of Frankfurters. International Journal of Food Science and Technology. 2009;**44**(6):1093-1099. DOI: 10.1111/j.1365-2621.2009.01921.x

[38] Kabel MA, Carvalheiro F, Garrote G, Avgerinos E, Koukios E, Parajo JC, Gírio FM, Schols HA, Voragen AGJ. Hydrothermally treated xylan rich by-products yield different classes of xylo-oligosaccharides. Carbohydrate Polymers. 2002;**50**(1):47-56. DOI: 10.1016/S0144-8617(02)00045-0

[39] Mussatto SI, Roberto IC. Acid hydrolysis and fermentation of brewer's spent grain to produce xylitol. Journal of the Science of Food and Agriculture. 2005;**85**(14):2453-2460. DOI: 10.1002/jsfa.2276

[40] Nagy M, Semeniuc CA, Socaci SA, Pop CA, Rotar AM, Sălăgean CD, Tofană M. Utilization of brewer's spent grain and mushrooms in fortification of smoked sausages. Food Science and Technology (Campinas). 2016. DOI: 10.1590/1678-457x.23816

[41] Stojceska V, Ainsworth P, Plunkett A, Ibanoglu S. The recycling of brewer's processing by-product into ready-to-eat snacks using extrusion technology. Journal of Cereal Science. 2008;**47**(3):469-479. DOI: 10.1016/j.jcs.2007.05.016

[42] Maneju H, Udobi CE, Joel N. Effect of added brewers dry grain on the physicochemical, microbial and sensory quality of wheat bread. American Journal of Food Nutrition. 2011;**1**(1):39-43. DOI: 10.5251/ajfn.2011.1.1.39.43

[43] Ktenioudaki A, Crofton E, Scannell GM, Hannon JA, Kilcawley KN, Gallagher E. Sensory properties and aromatic composition of baked snacks containing brewer's spent grain. Journal of Cereal Science. 2013;**57**(3):384-390. DOI: 10.1016/j.jcs.2013.01.009

[44] Klemm D, Kramer F, Moritz S, Lindström T, Ankerfors M, Gray D, Dorris A. Nanocelluloses: A New Family of Nature-Based Materials. Green Nanomaterials. 2011;**50**(73):5438-5466. DOI: 10.1002/anie.201001273

[45] Berglund L, Noël M, Aitomäki Y, Öman T, Oksman K. Production potential of cellulose nanofibers from industrial residues: Efficiency and nanofiber characteristics. Industrial Crops and Products. 2016;**92**:84-92. DOI: 10.1016/j.indcrop.2016.08.003

[46] Mishra PK, Wimmer R, Gregor T. Utilising brewer's spent grain as a source of cellulose nanofibres following separation of protein-based biomass. Bioresources. 2017;**12**(1):107-116. DOI: 10.15376/biores.12.1.107-116

[47] Mussatto SI, Fernandes M, Dragone G, Mancilha IM, Roberto IC. Brewer's spent grain as raw material for lactic acid production by Lactobacillus delbrueckii. Biotechnological Letters. 2007;**29**(12):1973-1976. DOI: 10.1007/s10529-007-9494-3

[48] Cruz JM, Moldes AB, Bustos G, Torrado A, Domínguez JM. Integral utilisation of barley husk for the production of food additives. Journal of the Science of the Food and Agriculture. 2007;**87**(6):1000-1008. DOI: 10.1002/jsfa.279

[49] Barbosa-Pereira L, Bilbao A, Vilches P, Angulo I, Luis J, Fité B, Paseiro-Losada P, Cruz JM. Brewery waste as a potential source of phenolic compounds: Optimization of the extraction process and evaluation of antioxidant and antimicrobial activities. Food Chemistry. 2014;**145**:191-197. DOI: 10.1016/j.foodchem.2013.08.033

[50] Meneses NGT, Martins S, Teixeira JA, Mussatto SI. Influence of extraction solvents on the recovery of antioxidant phenolic compounds from brewer's spent grains. Separation and Purification Technology. 2013;**108**:152-158. DOI: 10.1016/j.seppur.2013.02.015

[51] McCarthy AL, O'Callaghan YC, Connolly A, Piggott CO, FitzGerald RJ, O'Brien NM. A study of the ability of bioactive extracts from brewers' spent grain to enhance the antioxidant and immunomodulatory potential of food formulations following in vitro digestion. International Journal of Food Science and Nutrition. 2015;**66**(2):230-235. DOI: 10.3109/09637486.2014.979314

[52] Tatullo M, Simone GM, Tarullo F, Irlandese G, Vito D, Marrelli M, Santacroce L, Cocco T, Ballini A, Scacco S. Antioxidant and antitumor activity of a bioactive polyphenolic fraction isolated from the brewing process. 2016;**6**:36042. DOI: 10.1038/srep36042

[53] Spinelli Sara, Conte A, Del Nobile MA. Microencapsulation of extracted bioactive compounds from brewer's spent grain to enrich fish-burgers. Food and Bioproducts Processing. 2016;**100**(A):450-456. DOI: 10.1016/j.fbp.2016.09.005

[54] Becker R. Fatty acids in food cereal grains and grain products. In: Chow CK, editors. Fatty Acids in Foods and Their Health Implications, 3rd ed. London, UK: CRC Press; 2007, pp. 303-316

[55] Niemi P, Tamminen T, Smeds A, Viljanen K, Ohra-Aho T, Holopainen-Mantila U, Faulds CB, Poutanen K, Buchert J. Characterization of lipids and lignans in brewer's spent grain and its enzymatically extracted fraction. Journal of Agricultural and Food Chemistry. 2012;**60**(39):9910-9917. DOI: 10.1021/jf302684x

[56] Stanislava G. Barley grain non-specific lipid-transfer proteins (ns-LTPs) in beer production and quality. Journal of the Institute of Brewing. 2007;**113**(3):310-324. DOI: 10.1002/j.2050-0416.2007.tb00291.x

[57] Faulds CB, Robertson JA, Waldron KW. Effect of pH on the solubilization of brewers' spent grain by microbial carbohydrases and proteases. Journal of Agricultural and Food Chemistry. 2008;**56**(16):7038-7043. DOI: 10.1021/jf800433c

[58] Dubois V, Breton S, Linder M, Fanni J, Parmentier M. Fatty acid profiles of 80 vegetable oils with regard to their nutritional potential. European Journal of Lipid Science and Technology. 2007;**109**(7):710-732. DOI 10.1002/ejlt.200700040

[59] Bleoanca I, Bahrim G. Overview on brewing yeast stress factors. Romanian Biotechnological Letters. 2013;**18**(5):8559-8572

[60] Hellborg L, Piscur J. Yeast diversity in the brewing industry. In: Preddy VR, editors. Beer in Health and Disease Prevention. New York, US: Elsevier; 2009, pp. 1068-1073

[61] The Brewers of Europe. Guidance Note for Establishing BAT in the Brewing Industry. 2002. Available from: http://www.cerveceros.org/pdf/cbmcguidance-note.pdf [Accessed: 2017.02.16]

[62] Bokulich NA, Bamforth CW. The microbiology of malting and brewing. Microbiology and Molecular Biology Reviews. 2013;**77**(2):157-172. DOI: 10.1128/MMBR.00060-12

[63] Ferreira IMPLVO, Pinho O, Vieira E, Tavarela JG. Brewer's Saccharomyces yeast biomass: characteristics and potential applications. Trends in Food Science and Technology. 2010;**21**(2):77-84. DOI: 10.1016/j.tifs.2009.10.008

[64] Mathias TRS, Alexandre VMF, Cammarota MC, de Mello PPM, Sérvulo EFC. Characterization and determination of brewer's solid wastes composition. Journal of the Institute of Brewing. 2015;**121**(3):400-404. DOI: 10.1002/jib.229

[65] Amorim M, Pereira JO, Gomes D, Pereira CD, Pinheiro H, Pintado M. Nutritional ingredients from spent brewer's yeast obtained by hydrolysis and selective membrane filtration integrated in a pilot process. Journal of Food Engineering. 2016;**185**:42-47

[66] Vieira EF, Carvalho J, Pinto E, Cunha S, Almeida AA, Ferreira IMPLVO. Nutritive value, antioxidant activity and phenolic compounds profile of brewer's spent yeast extract. Journal of Food Composition and Analysis. 2016;**52**:44-51. DOI: 10.1016/j.jfca.2016.07.006

[67] Martins ZE, Erben M, Gallardo AE, Silva R, Barbosa Ines, Pinho O, Ferreira IMPLVO. Effect of spent yeast fortification on physical parameters, volatiles and sensorial characteristics of home-made bread. International Journal of Food Science and Technology. 2015;**50**(8):1855-1863. DOI: 10.1111/ijfs.12818

[68] Yoon TJ, Kim TJ, Lee H. Anti-tumor metastatic activity of β-glucan purified from mutated Saccharomyces cerevisiae. International Immunopharmacology. 2008;**8**(1):36-42. DOI: 10.1016/j.intimp.2007.10.005

[69] Petravic-Tominac V, Zechner-Krpan V, Berkovic K, Galovic P, Herceg Z, Srecec S, Spoljaric I. Rheological properties, water-holding and oil-binding capacities of particulate b-glucans isolated from spent brewer's yeast by three different procedures. Food Technology and Biotechnology. 2011;**49**(1):56-64

[70] Da Silva Araujo VB, Ferreira de Melo AN, Gaspar Costa A, Castro-Gomez RH, Madruga MS, de Souza EL, Magnani M. Followed extraction of β-glucan and mannoprotein from spent brewer's yeast (Saccharomyces uvarum) and application of the obtained mannoprotein as a stabilizer in mayonnaise. Innovative Food Science and Emerging Technologies. 2014;**23**:164-170. DOI: 10.1016/j.ifset.2013.12.013

[71] Bohn JA, BeMiller JN. (1-3)-β-D-Glucans as biological response modifiers: a review of structure-functional activity relationships. Carbohydrate Polymers. 1995;**28**(1):3-14. DOI: 10.1016/0144-8617(95)00076-3

[72] Waszkiewicz-Robak B. Spent brewer's yeast and beta-glucans isolated from them as diet components modifying blood lipid metabolism disturbed by an atherogenic diet. In: Valenzuela Baez R, editor. Lipid Metabolism. Rijeka, Croatia: InTech; 2013, pp. 262-290. DOI: 10.5772/51530

[73] EFSA Panel on Dietetic Products, Nutrition and Allergies (NDA). Scientific opinion on the safety of 'yeast beta-glucans' as a novel food ingredient. EFSA Journal. 2011;**9**(5):2137-2159. DOI: 10.2903/j.efsa.2011.2137

[74] Zechner-Krpan V, Petravic-Tominac V, Panjkota-Krbavcic I, Grba S, Berkovic K. Potential application of yeast β-glucans in food industry. Agriculturae Conspectus Scientificus. 2009;**74**(4):277-282

[75] Borchani C, Fonteyn F, Jamin G, Destain J, Willems L, Paquot M, Blecker C, Thonart P. Structural characterization, technological functionality, and physiological aspects of fungal β-d-glucans: A review. Critical Reviews in Food Science and Nutrition. 2016;**56**(10):1746-1752. DOI: 10.1080/10408398.2013.854733

[76] Rizzo M, Ventrice D, Varone MA, Sidari R, Caridi A. HPLC determination of phenolics adsorbed on yeasts. Journal of Pharmaceutical and Biomedical Analysis. 2006;**42**(1):46-55. DOI: 10.1016/j.jpba.2006.02.058

[77] Podpora B, Swiderski F, Sadowska A, Piotrowska A, Rakowska R. Spent brewer's yeast autolysates as a new and valuable component of functional food and dietary supplements. Journal of Food Processing and Technology. 2015;**6**:526. DOI: 10.4172/2157-7110.1000526

[78] Harlow BE, Bryant RW, Cohen SD, O'Connell SP, Flythe MD. Degradation of spent craft brewer's yeast by caprine rumen hyper ammonia-producing bacteria. Letters in Applied Microbiology. 2016;**63**(4):307-312. DOI: 10.1111/lam.12623

[79] Bryant RWR, Cohen SD. Characterization of hop acids in spent brewer's yeast from craft and multinational sources. Journal of the American Society of Brewing Chemists. 2015;**73**(2):159-164. DOI: 10.1094/ASBCJ-2015-0315-01

[80] Blagovic B, Rupcic J, Mesaric M, Katica Georgiu K, Maric V. Lipid composition of brewer's yeast. Food Technology and Biotechnology. 2001;**39**(3):175-181

[81] Bekatorou A, Psarianos C, Kotinas AA. Production of food grade yeasts. Food Technology and Biotechnology. 2006;**44**(3):407-415

[82] Podpora B, Swiderski F, Sadowska A, Rakovska R, Wasiak-Zys G. Spent brewer's yeast extracts as a new component of functional food. Czech Journal of Food Sciences. 2016; **34**(6):554-563. DOI: 10.17221/419/2015-CJFS

[83] Chae HJ, Joo H, Man-Jin I. Utilization of brewer's yeast cells for the production of food-grade yeast extract. Part 1: Effects of different enzymatic treatments on solid and protein recovery and flavor characteristics. Bioresource Technology. 2001;**76**(3):253-258. DOI: 10.1016/S0960-8524(00)00102-4

[84] Vieira E, Brandao T, Ferreira IMPLVO. Evaluation of brewer's spent yeast to produce flavor enhancer nucleotides: Influence of serial repitching. Journal of Agricultural and Food Chemistry. 2013;**61**(37):8724-8729. DOI: 10.1021/jf4021619

[85] Kuroda M, Myamura N. Mechanism of the perception of "kokumi" substances and the sensory characteristics of the "kokumi" peptide, γ-Glu-Val-Gly. Flavour. 2015;**4**:11. DOI: 10.1186/2044-7248-4-11

[86] Ruckle L, Senn T. Hop acids can efficiently replace antibiotics in ethanol production. International Sugar Journal. 2006;**108**(1287):139-147

Section 5

Quality Control of Beer

Chapter 8

Traditional Processing and Quality Control of the "Red Kapsiki": A Local Sorghum Beer from Northern Cameroon

Bayoï James Ronald and Djoulde Darman Roger

Additional information is available at the end of the chapter

http://dx.doi.org/10.5772/intechopen.69595

Abstract

This study was propose to elucidate the traditional process production, biochemical, and microbiological parameters of the "red kapsiki" beer locally called "*Te*." Direct interviews are conducted on the basis of questionnaires in four localities of the Far-North region of Cameroon. At each site, beer samples are collected, labeled, and undergo physicochemical and microbiological analyses using standardized methods. The results show that the traditional "red kapsiki" beer process incorporates a malting step, a large brewing stage, and a final fermentation step which requires a starter. The biochemical parameters of the beer samples show a pH value ranging from 2.40 ± 0.19 to 3.26 ± 0.03 (pH < 4.5), an alcohol content between 3.85 ± 0.58 and 4.28 ± 0.78% (v/v), a soluble extract which varies from 6.30 ± 1.09 to 7.29 ± 0.26 °P, a total sugar content which fluctuates between 41.8 ± 0.39 and 72.9 ± 0.40 g/L. In addition, the "red kapsiki" beer shows a total polyphenol content between 843 ± 27 and 1150 ± 27 mg/L and a flavonol level fluctuating between 750 ± 23 and 1300 ± 27 mg/L. Microbial analyses show a poor hygienic quality according to Cameroon standards referenced on the French Agency Norms.

Keywords: Cameroon, beer, homemade, processing, biochemistry, microbial, quality

1. Introduction

The artisanal fermentation of cereals into beers and wine-like alcoholic drink is not recent in Africa. The traces of the first artisanal fermentation were found by archaeologists in the

Blue Nile region of Sudan [1, 2]. In mountainous area of central African savannah, cereals, mainly millet and sorghum, are the most important crops used for fermented beverage [3]. One of these drinks is an opaque alcoholic beer-like beverage made from fermented sorghum gritz and malted maize, mainly used for rituals and festivities [4]. Two types of this beverage are produce in "Kapsiki land." They are "tè" and "mpedli." The first one, "tè," also called "red kapsiki," is the ritual beer. And the second, "mpedli," is the "white kapsiki" beer which is mainly brewed by women for commercial purpose and has no ritual significance. "Mpedli" is made by a quick process for immediate consumption [5, 6]. The red "tè" beer in which we focus our study is traditionally a man's brew. Its processing follows a strict procedure, with numerous prohibitions, and "red kapsiki" was for long time mainly used for ritual purposes than festivities. Symbolism was more focused on this beer rather than commercial and technical [6]. However, "red kapsiki" has increasingly become a sales commodity for women both at the village markets and in the cities as it is generally preferred by the population over "mpedli" [7, 8]. Though most of the symbolism around beer is a male-dominated discourse which concentrates on bonding and power, the symbolism is less straight forward and more hidden [6]. Despite the importance of the "red kapsiki" beer among this tribe, the beer itself remains unknown in scientific community and little is known on it processing and quality. This chapter thus aims at valorizing this opaque beer by describing its processing technique and provides some data in regard with its quality.

2. Materials and methods

2.1. Field work and sampling

In order to describe and follow the process production, a survey was conducted in three "kapsiki" rural villages of Cameroon, namely, Rhumsiki, Rhumzu, and Mogodé. Later on, some samples from one urban town close to the area Mokolo were also collected for comparison purpose. The choice of this urban area is justified by the fact that Mokolo is the immediate administrative area populated in majority by "kapsiki" populations. The sampling method used for processing description was cluster sampling [9] with two degrees of units. At the primary level, we have considered the cities surveyed, production sites and markets. At the secondary level, we have interviewed individuals and groups of individual respondents. Two layers were formed: rural area and urban area. As urban area, Mokolo was chosen because of the possibility of finding markets as well as production sites. In mountainous "Kapsiki" land, three villages were selected: Rhumsiki, Rhumzu, and Mogodé. As for the surveyed markets, we conducted a random choice in Mokolo. In the rural areas, Rhumsiki, Rhumzu, and Mogodé, all markets were taken into account because of the very limited number of markets and their periodical characters. Producers and women retailers were also randomly chosen in the areas and markets selected for the survey. A total of 15 production sites and 7 markets were visited, and 50 producers and 23 women retailers were interviewed. The interviews were conducted on the basis of a questionnaire, and collected data were processed using the software Winstat through a counting sheet constructed from the questionnaire.

2.2. On-site experiment

For characterization, 40 samples of the "red kapsiki" were collected from production sites and sales points. The sample pH was measured directly onsite using a portable pH meter. The conductivity, density, and brix were also recorded onsite using portables devices conductometer, densitometer, and brix meter, respectively. Around 10 ml of each sample was introduced into test tubes and gently shaken. The probe of designated apparatus (pH meter, conductometer, and densitometer) was then deepening into the test tube, and the values were read directly on the screen of the device. The experiment was repeated four times for each sample. The mean of each read result was considered. For total soluble solids (% Brix), the refractometric method was used to determine the soluble solids in beer samples [10]. The portable refractometer was first thermostated at 20°C using boiled water and regularly calibrated with cooled distilled water until the screen of the device showed 20°C. Soluble solids were then obtained from read refractive index on device screen, by reference to a standard table.

2.3. Laboratory experiment

2.3.1. Physicochemical analysis

2.3.1.1. Titrable acidity

Titrable acidity (as percentage (w/w) tartaric acid) was determined according to the Association of Analytical Chemists' method [11]. Acidity was evaluated by the alkali-potentiometric method using a 0.1-N sodium hydroxide solution in the presence of 0.4% bromothymol blue used as an indicator.

2.3.1.2. Total polyphenols

Total polyphenols were assayed by calorimeter using the Folin-Dennis Ciocalteau reagent as described by Mangas et al. [12], and the results were expressed as mg/l of gallic acid.

2.3.1.3. Total ethanol

Total ethanol content was preceded by a Spectrophotometric micro-method for the determination of ethanol after distillation of beer that was made alkaline by a suspension of calcium hydroxide [13].

2.3.1.4. Specific density at 15°C

Specific density at 15°C was evaluated as described by Nanda et al. [14]. The specific gravity determination was done as follows: 20 ml of sample was poured into the specific gravity test tube to overflow, then the stopper was inserted, and the whole was incubated in water bath at 200°C for 30 minutes. The test tube was removed from the water bath, wiped, and then weighed. Thereafter, the sample was boiled. The specific gravity was calculated as the ratio of weight of ash over the weight of fresh sample time 100.

2.3.1.5. Volatile acidity

The volatile acidity was determined using the Mathieu method by titration of the volatile acids separated from wine by steam distillation and titration of the distillate [15].

2.3.2. Microbial analysis

Around 10 ml of "red kapsiki" samples from each site was mixed with 90-ml sterile peptone physiological saline solution (1 g Peptone, 8.5 g NaCl, and 1000-ml distilled water). Decimal dilutions were prepared up to 10^6 from initial sample as described by Loyer and Hamilton [16]. All enumeration in solid media was carried out in triplicate, and the plates containing between 33 and 333 colonies were considered. The enumeration in liquid media was evaluated according to deMan most probable number.

2.3.2.1. Total aerobic mesophilic bacteria

Total aerobic mesophilic bacteria were enumerated on Plate Count Agar (PCA-OXOID) supplemented with cycloheximide 0.5% [17]. The plates were incubated at 28°C for 48–72 hours.

2.3.2.2. Total coliforms and Escherichia coli

Total coliforms and *Escherichia coli* were accessed on Bubble Lactose Bile with Brilliant green (BLBVB- DIFCO). The tubes containing the Durham bells were incubated at 30°C during 24–48 hours. The positives tubes were used to inoculate another test tube containing water peptone without indole and were incubated at 44°C for 24 hours for *E. coli* determination, which was revealed using Kovac's reagent [18].

2.3.2.3. Streptococcus

Fecal Streptococcus was enumerated on Slant Agar (SL-Merck) supplemented with cycloheximide at 0.5% after 48 hours of incubation at 37°C [19].

2.3.2.4. Salmonella and Shigella, yeasts and molds, sulfite-reducing

- *Salmonella* and *Shigella* were analyzed as described by Ribot et al. [20].
- Yeasts and molds were enumerated on PDA-Chloramphenicol (200 g potatoes extract, 10 g peptone, 20 g glucose, 15 g agar, 0.5 g chloramphenicol, and 1000 ml distilled water, pH was adjusted to 5.2) after 48–72 hours of incubation at 30°C [18].
- Enumeration of sulfite-reducing clostridia was done according to Mossel [21] method in anaerobic jar.

2.3.2.5. Total spore-forming bacteria

Total spore-forming bacteria were evaluated on GPB agar medium (10-g peptone, 2-g starch, 5-g glucose, 15-g agar, 40-mg bromocresol purple, and 1000-ml distilled water) after 10 minutes of pre-heating of the samples at 80°C [22]. Colony counts were performed after 48 hours of incubation at 35°C.

2.4. Statistical analysis

Comparison of the means was performed by the ANOVA associated with Tukey's honest significant difference (HSD) test to discriminate significantly different pairs of means values. Means values were considered statistically different at $P \leq 0.05$ significance level.

3. Results and discussion

3.1. Traditional processing of the "red kapsiki"

Sorghum beer is generally made from grain and water, sometimes a gelatinous or mucilaginous agent [23]. In Cameroon, most of non-Islamized ethnic groups process it even in different forms. The obtained beer is named according to ethnic groups and countries. Names of beer are as follows: "Tchoukoutou" in Togo and Benin, "Pito" and "Burukutu" in Ghana and Nigeria, "Dolo" in Burkina-Faso, "Tchapalo" in Ivory Coast, "Busaa" and "Bushera" in Kenya, "Ikigage" in Rwanda, Kaffir in South Africa, "Mahewu" in Zimbabwe, "Malwa" in Uganda, "Munkoyo" in DR Congo, and "Bili-Bili" in Cameroon. **Figure 1** describes the traditional processing of "red kapsiki" beer which commonly includes the following:

3.1.1. Malting

In the case of "red kapsiki," as most of beers, the process production starts by the selection of grains. The "red kapsiki" being a noble beer, only good quality grains are considered. Mostly, the sorghum variety "mouskwari" is selected for the "red kapsiki" production. However, in raining season, "Djigari" variety can also be chosen.

3.1.1.1. Quenching

After being washed, the grains are immersed in water for 12–24 hours so as to obtain a moisture content of 35–40% for germination. The temperature of water is very important: at high temperature, the quenching is rapid. The immersion temperature is close to that of room temperature (around 40–45°C in the region). The grains are first drained on tissue and then stabled in double layer on cotton cloth bags or on woven mats.

3.1.1.2. Germination

The soft grains are covered and placed in dark area for 2–3 days for germination. Water is sprayed sometimes, when the ambient air is dry or when the temperature is hot. Alternatively,

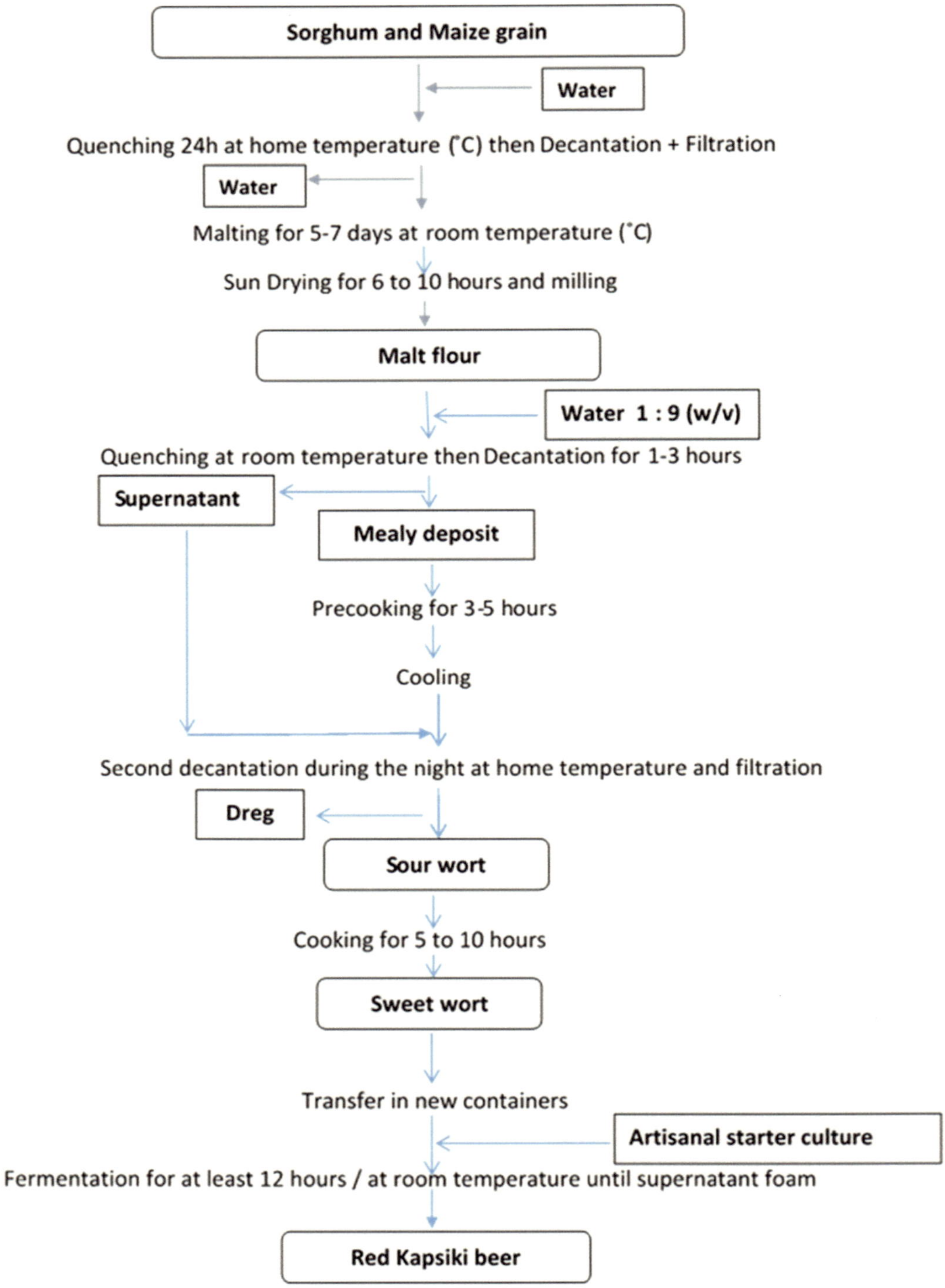

Figure 1. Technological diagram of homemade "red kapsiki" beer.

the grains are left on the ground and sprayed until the germination process starts and rootlets appear. The high temperature facilitates the beginning of germination. In this case, the germination time can last for 4 days. It should be noted that the same technique is used at the household level to improve the energy density of slurries [24]. During germination, amylolytic enzymes are produced and protein digestibility of sorghum, which is generally low, is improved [25]. It was also demonstrated that after 3 days of malting, there was production of amylolytic enzymes, including α-amylase, β-amylase, and dextrinase, which are all essential for good quality of malt [26].

3.1.1.3. Drying

This corresponds to the "Kilning" and brings moisture to malt to keep 15–20% without mold. The malt is dried in sun for one or more days, sometimes less if the process goes straight to the brewing stage. In case of the production of special "red kapsiki," it was noticed that after malting, grains are roasted in firewood and ground coarsely. The obtained powder is kept in dark for 2 days before brewing.

3.1.2. Brewing

3.1.2.1. Milling

The previously malt is crushed in a mortar/pestle in rural sites or in a wen in urban area.

In fact, the malt is brought to a motorized milling machine sets to crushing mode, to obtain a coarse flour.

3.1.2.2. Pasting

The grind is mixed with water and a gelatinous or mucilaginous agent (okra or sap of various trees that improve flocculation and filtration of insoluble in suspension). After an hour of storage at temperature of 25 to 35°C, the mixture was separated into two phases, the upper liquid phase which is collected. The upper liquid phase already contains a soluble portion of malt sugar.

3.1.2.3. The decoction

The lower phase containing malt flour is cooked slowly to boiling so as to obtain a starch paste (slurry consistency). The upper liquid base is then mixed with a water to be more easily saccharified than if it was not being cooked, and the diastatic actions being more effective on cooked starch than on raw starch. Alternatively, like for other sorghum beer processing, where the raw grain is added [27, 28, 29], in the case of the "red kapsiki," production may continue with malted sorghum powder. In fact, ground malted sorghum is dissolved in water at the ratio of 1/9 (w/v). After 1–3 hours of soaking, supernatant is removed and kept for a later use. The remaining mealy material at the bottom of the soaking container is then removed and cooked for 3–5 hours. It must be noticed here that some enzymes produced during the malting stage seem not digested. This may be due to the soaking temperature, which is not optimal for enzymes. The mealy deposit is constituted by 80% raw starch. This starch is cooked and lightly cooled before previously removed soaking water probably containing starch-digesting enzymes is added. The mixture is then kept warm for 1–5 hours or let stand overnight at room temperature.

3.1.2.4. Filtration

After the decoction phase, the paste-like mixture previously removed becomes liquid and is filtrated. The dry matter is discarded, while the sour mash obtained is kept for the next step of the process. The filtration is mostly done through polypropylene bags, and the slurry obtained after decoction is passed through this polypropylene bag. The filtrate now called "liquid must" is kept for further processing when the drench is used as animal feed.

3.1.3. Cooking

The liquid must be concentrated and clarified by skimming. This operation is stopped by several criteria: clarity, color of the must, cold consistency (syrupy appearance) and also the flavor of the must. This operation consisted of two cooking steps. The first step is a precooking of mealy deposit during 3–5 hours to produce a "sour liquid must." The second one is a cooking of the "sour liquid must" 5–10 hours to produce a "sweetish liquid must" which is called "tè kwarhèni" in the local Kapsiki dialect.

3.1.4. Fermentation

The sweetish must is cooled either spontaneously or by successive decanting and then starter culture is added. Fermentation lasts 12–24 hours at room temperature. This last step is stopped when supernatant liquid foam. The beverage obtained at the end is called "tè" or "red kapsiki."

3.2. Biochemical profile of the "red kapsiki"

As presented in **Table 1** and compared to other African beer as described by Lyumugabe et al. [29], the "red kapsiki" presents a greater alcohol content (3.85–4.28% v/v). This beer seems more alcoholic than "Bushera" 0.27% [30], "Burukutu" 1.63% [31], "Munkoyo" 2.1% [32], "Dolo" 2.3% [33], "white Kapsiki" beer (2.48 ± 0.14%) [34], and "Pito" (3.09%) [35]. However, the "red kapsiki" would be less alcoholic than "Tchapalo" (5.03–5.22%) [28]. We noticed a pH between 2.40 ± 0.19 and 3.26 ± 0.03. With an average pH below 4.5, the "red kapsiki" beer samples would be of satisfactory quality according to the CODEX STAN 243. Soluble extract varies from 6.30 to 7.29 °P,Brix from 7.0 to 7.46 °B, and total sugar from 41.8 to 72.9 g/l. Compared with the literature, this beer seems to be sweeter than "Tchapalo" 5.3 g/l [28], "Dolo" 7.7 g/l [33], and "Pito" 34 g/l [35]. This beer has a conductivity from 1919 to 1990 (μS/cm) and a specific density (g/cm) at 15°C of about 1.33. The color of the "red kapsiki" varies from a pinky brown to reddish according the variety of sorghum used. As most of African sorghum beers, the "red kapsiki" presents a touch of fruitiness added to their fermentation odor. This beer is mainly consumed

http://dx.doi.org/10.5772/intechopen.69595

	Mogodé	Mokolo	Rhumsiki	Rhumzu
pH	2.46 ± 0.08[a]	2.42 ± 0.12[a]	2.40 ± 0.19[a]	3.26 ± 0.03[a]
Total titrable acidity (mg/l)	6.7 ± 0.4[a]	8.1 ± 0.5[b]	7.7 ± 0.1[b,c]	7.2 ± 0.6[a,c]
Soluble extract (°P)	7.28 ± 1.29[a]	7.29 ± 0.26[a]	7.29 ± 0.26[a]	6.30 ± 1.09[a]
Brix (°B)	7.0 ± 1.06[b]	7.46 ± 0.83[b]	7.42 ± 0.84[b]	7.0 ± 0.16[b]
Total ethanol (% vol)	3.85 ± 0.58[a]	4.10 ± 0.46[a]	4.08 ± 0.46[a]	4.28 ± 0.78[a]
Total sugars (g/l)	72.8 ± 1.29[a]	72.9 ± 0.30[a]	72.9 ± 0.40[a]	41.8 ± 0.39[a]
Conductivity (μS/cm)	1919.23 ± 8.12[a]	1990.0 ± 4.08[b]	1990.0 ± 3.53[b]	1929.00 ± 4.02[a]
Specific density (g/cm) 15°C	1.03 ± 0.00[a]	1.33 ± 0.00[a]	1.00 ± 0.00[a]	1.62 ± 0.00[a]

NOTE: Mean values preceded by at least one common letter (a, b, c) in the same line are not significantly different (P <0.05) according to the ANOVA and Tukey comparison test.

Table 1. Biochemical profile of the "red kapsiki" beer.

in an actively fermenting state leading to a short shelf life as mentioned for other African beer in literature [20, 29, 36, 37]. Statistical analyses carried out on the physicochemical composition of "kapsiki red" beers show generally that there is no significant difference between the samples from the different sites. Indeed, the physicochemical parameters of this beer are substantially the same from one site to another. Despite a difference in the manufacturing process, the results obtained with this beer are similar to those of Yao et al. [38], which showed a consistency in the physicochemical properties of "Tchapalo" taken from nine different sites in the city of Abidjan. At first glance, this regularity seems surprising since the artisanal production of "kapsiki red" beer is made without measuring and precision equipment. The operations are done by simple visual and sensorial appreciation [28]. This apparent invariability could be explained by the fact that manufacturing being empirical, brewers have kept the same reflexes and habits. This allows them to obtain more or less identical finished products.

As present in **Table 2**, the "red kapsiki" contains a quite good amount of polyphenols. The recorded amount varies from 843 ± 27 mg/l in Mogodé samples to 1150 ± 27 mg/l in Rhumzu samples. It must be noticed that some of these polyphenols are too low or absent in other "industrial" beer. As indicated by Bröhan et al. [39], when barley malt is used for mashing, around 30% of total beer polyphenols are issued from hop, although added in 100 times lesser quantity than malt. In the case of the "te" or "red kapsiki," the sorghum contribution to beer polyphenols could be much higher. In fact, sorghum phenolic acids include hydroxybenzoic (mainly protocatechuic acid) and hydroxycinnamic acids [40, 41] both free and bound as esters. Most of them are found in usual lager beers brewed either from barley malt or from hop [42]. Sorghum anthocyanins are unique, as they lack the hydroxyl group at the 3-position of the C ring. These 3-deoxyanthocyanins such as luteolinidin and apigeninidin are used as natural food colorings because they are more stable than anthocyanidins in both organic solvents and acidic solutions. Amount of 1500 mg/l of flavonol was recorded in "red kapsiki." Bröhan et al. [39] indicate that flavonols such as apiforol (leucoapigeninidin) and luteoforol (leucoluteolinidin) are sorghum polyphenols as precursors of sorghum 3-deoxyanthocyanins.

Sample	Volatile acidity (g/l)	Total polyphenols (mg/l)	Flavonol (mg/l)
Rhumzu	0.3 ± 0.00[a]	1150 ± 27[a]	1300 ± 27[a]
Rhumsiki	0.2 ± 0.00[b,c]	911 ± 22[b]	1000 ± 32[c]
Mogodé	0.1 ± 0.00[c]	843 ± 27[b,a]	834 ± 16[a,d]
Mokolo	0.2 ± 0.00[b,c]	1111 ± 32[c,a]	750 ± 23[e,a]

NOTE: Mean values preceded by at least one common letter (a, b, c, d, e) in the same line are not significantly different (P <0.05) according to the ANOVA and Tukey comparison test.

Table 2. Some essential biochemical compounds of the "red kapsiki".

Never reported in beer, they have been found at concentrations up to 4200 mg/kg in sorghum [43]. Other sorghum flavonoids include the flavones apigenin and luteolin [44], the flavanones naringenin and eriodictyol [45], the flavonol kaempferol, the dihydroflavonol taxifolin, and the flavan-3-ols (þcatechin and epicatechin. Hop brings similar flavonols and flavan-3-ols to wort, in industrial brewing.

3.3. Microbial profile of the "red kapsiki"

The results of the microbiological analysis of beers obtained are shown in **Table 3**. Analysis of these results shows that the parameters sought in the produced beers are not in accordance with international microbiological criteria [46]. The presence of pathogens as Coliforms, *Salmonella* and *Shigella,* and yeasts and molds in these beverages indicates that the "red kapsiki" is of bad hygienic quality. In fact, it's said that one of the main factors limiting the use of the "red kapsiki" like most of African opaque beers is that they spoil rapidly due to extra bacterial action. Despite its low acidity and pH, the load of microorganisms is important (**Table 3**). Total aerobic microflora is up to 6.2 10^7 cfu/ml. This may be explained by the fact that the "red kapsiki" is still actively fermenting when sold and consumed. This means that the process is not optimal yet when the beverage is consumed. The total coliform, total spore-forming bacteria, and clostridia loads are higher in Mokolo sample than in other samples, with values of (2.4 ±0.7) 10^5cfu/ml, (7.7 ±0.3) 10^3 cfu/ml, and (3.9 ± 0.6)10^3 cfu/ml, respectively. The presence of coliforms and fecal Streptococcus genera is obvious as the "red kapsiki" presents a too low acidity level and presence of alcohol even in insufficient amount. It was expected to have a synergetic effect of acid and alcohol against microorganisms. Among the pathogenic microorganisms could be isolated in craft beers, we can mentioned coliforms as *Escherichia coli* and spore-forming bacteria species. Their presence and persistence in these beverages would not only be linked to a simple contamination but also to their adaptation ability. Indeed, several studies have shown that some environmental parameters such as low temperatures like those that are observed during processing have the capacity to induce the resistance of these microorganisms to high temperatures [47] and strongly acidic pH [22, 48]. Bayoï et al. [22] showed that spores of *Bacillus subtilis* and *Geobacillus stearothermophilus* pre-treated at 45, 50, and 60°C during 1–3 hours before treatment with acetic acid at pH 4.5 were significantly more resistant to this acid compared to the spores of the same bacterial species non-pretreated and subjected under the identical acid conditions. Etoa and Adegoke [47]

http://dx.doi.org/10.5772/intechopen.69595

	Mogodé	Mokolo	Rhumsiki	Rhumzu	Standards
Total count (cfu/ml)	$(6.1 \pm 0.2)\ 10^{5\,\mathrm{b}}$	$(6.2 \pm 0.5)\ 10^{7\,\mathrm{a}}$	$(5.1 \pm 0.3)\ 10^{5\,\mathrm{b}}$	$(7.4 \pm 0.1)\ 10^{4\,\mathrm{b}}$	$<10^{6}$
Total coliform (cfu/ml)	$(7.2 \pm 0.5)\ 10^{4\,\mathrm{b}}$	$(2.4 \pm 0.7)\ 10^{5\,\mathrm{a}}$	$(4.2 \pm 0.4)\ 10^{3\,\mathrm{b}}$	$(1.4 \pm 0.3)\ 10^{4\,\mathrm{b}}$	$<10^{3}$
Total thermo-tolerant coliforms (cfu/ml)	$(9.2 \pm 0.4)\ 10^{1\,\mathrm{a}}$	$(3.1 \pm 0.4)\ 10^{3\,\mathrm{b}}$	$(5.2 \pm 0.7)\ 10^{1\,\mathrm{a}}$	$(3.1 \pm 0.5)\ 10^{2\,\mathrm{a,c}}$	$<10^{2}$
Fecal Streptococcus (cfu/ml)	$(2.2 \pm 0.2)\ 10^{2\,\mathrm{c}}$	$(3.7 \pm 0.4)\ 10^{4\,\mathrm{a}}$	$(3.2 \pm 0.3)\ 10^{3\,\mathrm{c}}$	$(2.2 \pm 0.2)\ 10^{4\,\mathrm{b}}$	$<10^{3}$
Salmonella and *Shigella* (cfu/20 g)	$(8.1 \pm 0.7)\ 10^{1\,\mathrm{b}}$	$(4.5 \pm 0.2)\ 10^{3\,\mathrm{a}}$	$(9.2 \pm 0.7)\ 10^{2\,\mathrm{b}}$	$(7.2 \pm 0.5)\ 10^{3\,\mathrm{c}}$	Absence/20g
Sulfite-reducing clostridia (cfu/ml)	$(5.0 \pm 0.4)\ 10^{1\,\mathrm{b}}$	$(7.7 \pm 0.3)\ 10^{3\,\mathrm{a}}$	$(7.2 \pm 0.9)\ 10^{1\,\mathrm{b}}$	$(2.5 \pm 0.6)\ 10^{2\,\mathrm{b}}$	Not known
Total fungi (cfu/ml)	$(3.5 \pm 0.8)\ 10^{3\,\mathrm{b}}$	$(3.2 \pm 0.7)\ 10^{5\,\mathrm{a}}$	$(4.5 \pm 0.4)\ 10^{4\,\mathrm{b}}$	$(6.1 \pm 0.2)\ 10^{4\,\mathrm{b}}$	$<10^{5}$
Total spore-forming bacteria (cfu/ml)	$(2.7 \pm 0.2)\ 10^{2\,\mathrm{a}}$	$(3.9 \pm 0.6)\ 10^{3\,\mathrm{b}}$	$(2.4 \pm 0.2)\ 10^{2\,\mathrm{a}}$	$(9.7 \pm 0.2)\ 10^{2\,\mathrm{a}}$	$<10^{4}$

NOTE: Mean values preceded by at least one common letter (a, b, c) in the same line are not significantly different ($P < 0.05$) according to the ANOVA and Tukey comparison test.

Table 3. Microbial profile of the "red kapsiki".

and Bayoï et al. [22] explained this phenomenon during which there was a slight increase in resistances in the spore following their stay at sublethal temperatures (heat-induced resistance), by the structural modifications of the spores molecules, mainly those of the different tunics and of the inner membrane. According to these authors, these molecules would undergo changes in conformation in presence of sublethal temperatures, which would result in the reduction of the permeability with respect to chemicals. Wang and Doyle [48] showed that survival in a minimum glucose medium at pH 2.5 of certain strains of *E. coli* pretreated at 48°C for 10 minutes was 10–100 greater compared to strains nonpretreated. According to Small et al. [49], this difference in acid resistance of pretreated *E. coli* strains could be the result of either the difference in the expression of the rpoS gene or the amount of synthesized heat shock proteins. Wang and Doyle [48] have clearly shown that the induction of acid resistance by thermal shock implies the synthesis of new proteins. According to these authors, two proteins, namely one of 22 KDa and another of 15 kDa, were synthesized at the outer membrane during the heat shock (48°C/10 minutes). Indeed, these two proteins would be subunits of an alkyl hydroperoxide reductase involved probably in the transport of protons out of the cell [48].This phenomenon was called "heat-induced acid resistance" [22]. The presence of *E. coli* would also probably be related to the induction of other mechanisms associated with acid resistance in *E. coli*. Indeed, in *E. coli* more specific systems neutralize the protons entering the cell which contributes to increase the internal pH (pHi). Four systems, known as AR or "Acid Resistance," have been identified. The first, AR1 inhibited in the presence of glucose, is able to protect the cell in stationary phase, and it is dependent on RpoS and the CRP-cAMP complex [50]. The presence of glucose in the medium reveals three other systems, AR2 (Arginine decarboxylase), AR3 (Lysine decarboxylase), and AR4 (Glutamate decarboxylase). These systems respectively decarboxylate arginine, lysine, and glutamate in agmatine, cadaverine, and g-aminobutyrate (GABA), which are expelled into the outside environment. This allows the consumption of a proton (H+) and the release of CO_2. Thus, when the external pH is 2.5, these decarboxylations allow the increase of the internal pH of 3.6 to 4.2–4.7, and the inversion of the membrane potential, essential for the survival of *E. coli* [51].

Indeed, studies on traditional sorghum beer in West Africa show that these drinks are a complex biotope composed of several genera and species of microorganisms dominated by yeast [29, 52, 53]. Their prevalence is probably because they are added by inoculation of the traditional starter in the sweet wort to ensure alcoholic fermentation [38, 54]. During fermentation, we initially have a yeast growth which is accompanied by the production of ethanol after the logarithmic growth phase, and which continues during the stationary phase. It has been observed that during this time, very little or no increase in the number of contaminating organisms seems to occur [18]. However, the presence of yeast in beer would also be due to adaptation in response to the accumulation of ethanol produced. According to Dombek and Ingram [55], this adaptation was accompanied by a change in the composition of membrane lipids after accumulation of ethanol. This changes in the composition and structure membrane would allow the yeasts to escape at the solubilizing action of ethanol. The surviving yeasts can readily begin another post-adaptation growth cycle. The isolated pathogenic bacteria can originate from the environment, the raw material, and the equipment used. The hygienic quality of beer produced depends closely on the conditions of fermentation of the must. Indeed, the levels of total sugar and vitamin C are relatively high in beers obtained from the fermented mash. The recorded values for the "red kapsiki" are higher than those in the "Tchapalo" [28, 54] where the fermentation is carried out at a room temperature with a starter culture based on previous productions. It must also be noticed, to explain the prevalence of pathogens, that after few days of fermentation, the amount of yeasts decreases because of autolysis. With little or no competition from yeasts for the readily available nutrients, contaminating microorganisms increase rapidly in number and their metabolites may change and spoil the beer. Because of the relatively high temperature of the "red kapsiki" fermentation, these sequential events occur within a short-time period. This period does not usually exceed more than 3 days in summer or 5 days in winter before this spoilage occurs. The metabolic activities of mesophilic bacteria are primarily responsible for the spoilage. These bacteria, along with other undesirable bacteria, may produce acetic acid, volatile off-flavors, fruity odors, and pellicles which render the taste, odor, and texture of the beer unacceptable to consumers.

4. Conclusion

The color of the "red kapsiki" is pink brown to reddish according the variety of sorghum used. This artisanal beverage requires steps of malting, quenching, germination and "kilning,"

http://dx.doi.org/10.5772/intechopen.69595

decoction, filtration, boiling and sterilizing, cooling, sowing, and fermentation for its production. The "red kapsiki" presents an interesting physicochemical profile, but has a high degree of microbial contamination for consumption, and we noticed the presence of pathogens such as coliforms, *Salmonella, Shigella,* as well as some alteration flora as yeasts and molds. The potential of the beverage for sales and income adds to the fact that this beverage has symbolic value for local population and there is a need to improve its entire process production and hygienic quality. Thus, for the advanced knowledge about certain characteristics of this local beer, it appears necessary to include in future studies a hazard analysis critical control point (HACCP) in order to propose a better manufacturing technology to ensure the production of a good hygienic "red kapsiki" beer.

Acknowledgements

This work was done with the support of Institut de Recherches pour le Developpement (IRD), France, through JEAIRD funding to TABITAL research team of the University of Maroua, Cameroon, which partially funded the field work.

Author details

Bayoï James Ronald[1] and Djoulde Darman Roger[2*]

*Address all correspondence to: jabar982002@gmail.com

1 Department of Biological sciences, Faculty of Science, University of Maroua, Maroua, Cameroon

2 The Higher Institute of the Sahel, University of Maroua, Maroua, Cameroon

References

[1] Jolly E. Boire avec esprit: bière de mil et société dogon. Société d'ethnologie, Collection Sociétés Africaines, Université de Paris X, Nanterre, France, 2004;**18**: p.493

[2] Djitog D, Kourmarou M, Maoura N. Organisation des unités de transformation artisanale en zone de savanes: cas de la transformation du sorgho en bière locale bili-bili à Moundou au Tchad. In: Savanes africaines: des espaces en mutation, des acteurs face à de nouveaux défis. Actes du colloque, Garoua, Cameroun. Cirad-Prasac, 2003;**6:** p.55

[3] Roger D, Venassius L, Justin ENJ, Xavier EF. Processing of "Amgba": A sorghum-maize based beer, brewed in Cameroon. Journal of Brewing and Distilling. 2013;**4**(1):11-18

[4] Van Beek WEA, Kapsiki beer dynamics. In: Raimond C, Garine E, Langlois O, editors. Food resources and food choices in the Lake Chad basin, IRD, Paris; 2006, pp. 477-500

[5] Fischler C. Notes sur les fonctions sociales de l'alcool. In: Caro G, editor. De l'alcoolisme au bien boire. Paris: L'Harmattan. 1990. pp. 162-168

[6] Van Beek WEA. Kapsiki beer dynamics. In: Garine E, Langlois O, Raimond C, editors. Ressources vivrières et choix alimentaires dans le bassin du Lac Chad. Paris: IRD. 2006. pp. 477-500

[7] Chrétien JP. L'histoire de longue durée de la consommation alimentaire en Afrique. In: de Garine I, editor. Les changements des habitudes et des politiques alimentaires en Afrique: Aspects des sciences humaines, naturelles et sociales. Paris, Publisud. 1991. pp. 63-83

[8] Bérard L, Marchenay P. Les dimensions culturelles de la fermentation. Les Fermentations au Service des produits de terroir. INRA, Paris, 2005:13-28, ISSN: 1250-5218, ISBN: 2738012035

[9] Kalton G, editor. Introduction to survey sampling. Vol. 35. 1983

[10] Joslyn MA. Refractometric methods. In: Joslyn MA, editor. Methods in Food Analysis. Physical, Chemical, and Instrumental Methods of Analysis. New York, NY: Academic Press, ed. by M.A. Joslyn, (2nd ed.), 1970;**9**:239-266

[11] A.O.A.C (Association of Official Analytical Chemists). In Helrich K. (Ed.), Official methods of analysis. (15th ed.), Arlington, VA, USA, (1990)

[12] Mangas JJ, Rodríguez R, Suárez B, Picinelli A, Dapena E. Study of the phenolic profile of cider apple cultivars at maturity by multivariate techniques. Journal of Agricultural and Food Chemistry. 1999;**47**(10):4046-4052

[13] Magrí Andrea D, Magrí AD, Magri AL, Balestrieri F, Sacchini A, Marini D. Spectrophotometric micro-method for the determination of ethanol in commercial beverages. Fresenius' Journal of Analytical Chemistry. 1997;**357**(7):985-988

[14] Nanda V, Sarkar BC, Sharma HK, Bawa AS. Physico-chemical properties and estimation of mineral content in honey produced from different plants in Northern India. Journal of Food Composition and Analysis. 2003;**16**(5):613-619

[15] Ribereau-Gayon J, Peynaud E. Application à la vignification de levures métabolisant l'acide L-malique. Académie d'Agriculture de France, 1962;**18**:558-563

[16] Loyer MW, Hamilton MA. Interval estimation of the density of organisms using a serial-dilution experiment. Biometrics. 1984;**40**:907-916

[17] Smith LB, Fox TL, Busta FF. Comparison of a dry medium culture plate (Petrifilm SM plates) method to the aerobic plate count method for enumeration of mesophilic aerobic colony-forming units in fresh ground beef. Journal of Food Protection®. 1985;**48**(12):1044-1045

[18] Roger DD, Daoudou B, James B, Etoa F. Artisanal Production of "kuri" an honey made alcoholic beverage from Adamaoua Cameroon. Global Journal of Scientific Researches GJSR. 2014;**2**(3):65-70. E-ISSN: 2311-2332

[19] Baird RM, Lee WH. Media used in the detection and enumeration of Staphylococcus aureus. Progress in Industrial Microbiology. 1995;**34**:77-87

[20] Ribot EM, Fair MA, Gautom R, Cameron DN, Hunter SB, Swaminathan B, Barrett TJ. Standardization of pulsed-field gel electrophoresis protocols for the subtyping of Escherichia coli O157: H7, Salmonella, and Shigella for PulseNet. Foodbourne Pathogens & Disease. 2006;**3**(1):59-67

[21] Mossel DAA. Enumeration of sulphite-reducing clostridia occurring in foods. Journal of the Science of Food and Agriculture. 1959;**10**(12):662-669

[22] Bayoï JR, Djoulde DR, Kamga PB, Nyegue M, Voundi O, Bakary D, Etoa FX. Activity of acetic acid on Bacillus stearothermophilus and Bacillus subtilis spores after sub-lethal heat pretreatments. International Journal of Innovation and Scientific Research. 2014;**10**(2):570-575

[23] Roger DD. Deoxynivanol (DON) and fumonisins B1 (FB1) in artisanal sorghum opaque beer brewed in north Cameroon. African Journal of Microbiology Research. 2011;**5**(12):1565-1567

[24] Verster A. Techniques de maltage: Expériences au niveau des Pays. In: Trèche S, de Benoist B, Benbouzid D, Verster A, Delpeuch F, editors. L'alimentation de complément du jeune enfant (Alexandrie. Egypte; 1994), Paris, ORSTOM, Collection Colloques et Séminaires, 1995, 333-338, ISSN 07672896

[25] Makokha AO, Oniang'o R, Njoroge SM, Kinyanjui PK. Effect of malting on protein digestibility of some sorghum (Sorghum bicolor) varieties grown in Kenya. African Journal of Food, Agriculture, Nutrition and Development. 2002;**2**(2):59-66

[26] Ba K, Tine E, Destain J, Cissé N, Thonart P. Étude comparative des composés phénoliques, du pouvoir antioxydant de différentes variétés de sorgho sénégalais et des enzymes amylolytiques de leur malt. Biotechnologie, Agronomie, Société et Environnement. 2010;**14**(1):131-139

[27] Gessain M, de Lestrange MT. La bière de sorgho chez les Bassari. Social Science Information. 1987;**26**(3):633-648

[28] Aka S, Djeni NDT, Guessan KFN. Variabilité des propriétés physico-chimiques et dénombrement de la flore fermentaire du tchapalo, une bière traditionnelle de sorgho en Côte d'Ivoire. Afrique Science. 2008;**4**(2):274-286

[29] Lyumugabe F, Gros J, Nzungize J, Bajyana E, Thonart P. Characteristics of African traditional beers brewed with sorghum malt: A review. Base[En ligne]. 2012;**4**(16):509-530

[30] Muyanja CMK, Narvhus JA, Treimo J, Langsrud T. Isolation, characterization and identification of lactic acid bacteria from bushera: A Ugandan traditional fermented beverage. International Journal of Food Microbiology. 2003;**80**(3):201-210

[31] Eze VC, Eleke OI, Omeh YS. Microbiological and nutritional qualities of burukutu sold in mammy market Abakpa, Enugu State, Nigeria. American Journal of Food and Nutrition. 2011;**1**(3):141-146

[32] Lovelace CEA. Estimation of nutrient content of two fermented beverages from Zambia opaque maize beer and munkoyo, Handbook of indigenous fermented foods. New York, USA: Marcel Dekker; 1977. pp. 371-373

[33] Abdoul-latif FM, Bassolé IHN, Dicko MH. Proximate composition of traditional local sorghum beer dolo manufactured in Ouagadougou. African Journal of Biotechnology. 2013;**12**(13):1517-1522

[34] Bayoï JR, Darman DR, Etoa FX. Technologie de fabrication, propriétés physico-chimiques et microbiologiques de la bière « kapsiki blanche » produite dans les monts Mandara au Nord –Cameroun. Afrique Sciences. 2016;**12**(6):123-134

[35] Fadahunsi IF, Ogunbanwo ST, Fawole AO. Microbiological and nutritional assessment of burukutu and pito (indigenously fermented alcoholic beverages in West Africa) during storage. Natural Science. 2013;**11**(4):98-103

[36] Yao AK, Kadio N, Coulibaly A, Agbo ZG. In processing and industrial utilization of sorghum and related cereal in Africa. In: Proceeding of the OUA/STRCSAFGRAD Regional symposium. Ouagadougou, Burkina Faso; 1995. pp. 22-26

[37] Novellie L, De Schaepdrijver P. Modern developments in traditional African beers. Progress in industrial microbiology, 1986;**23**:73-157

[38] Maoura N, Pourquie J. Sorghum beer: Production, nutritional value and impact upon human health. In: Preedy VR, editor. Beer in Health Disease Prevention. Burlington, MA, USA: Elsevier Academic Press; 2009. pp. 53-60

[39] Bröhan M, Jerkovic V, Collin S. Potentiality of red sorghum for producing stilbenoid-enriched beers with high antioxidant activity. Journal of Agricultural and Food Chemistry. 2011;**59**(8):4088-4094

[40] Beta T, Rooney LW, Marovatsanga LT, Taylor JRN. Phenolic compounds and kernel characteristics of Zimbabwean sorghums. Journal of the Science of Food and Agriculture. 1999;**79**:1003-1010

[41] Hahn DH, Faubion JM, Rooney LW. Sorghum phenolic acids, their high performance liquid chromatography separation and their relation to fungal resistance. Cereal Chemistry. 1983;**60**:255-259

[42] Callemien D, Collin S. Structure, organoleptic properties, quantification methods and stability of phenolic compounds in beer. A review. Food Reviews International. 2010;**26**:1-84

[43] Dicko MH, Gruppen H, Traore AS, Van Berkel WJH, Voragen AGJ. Evaluation of the effect of germination on phenolic compounds and antioxidant activites in sorghum varieties. Journal of Agricultural and Food Chemistry. 2005;**53**:2581-2588

[44] Dykes L, Rooney LW. Review-Sorghum and millet phenols and antioxidants. Journal of Cereal Science. 2006;**44**:236-251

[45] Dykes L, Seitz LM, Rooney WL, Rooney LW. Flavonoid composition of red sorghum genotypes. Food Chemistry. 2009;**116**:313-317

[46] Legislative and regulatory guide French, December 2000, No. 815512

[47] Etoa FX, Adegoke GO. Mécanisme et implication de la thermorésistance de la spore bactérienne. Microbiologie et Hygiene Alimentaire. 1995;**18**(7):3-7

[48] Wang G, Doyle MP. Heat shock response enhances acid tolerance of Escherichia coli O157: H7. Letters in Applied Microbiology. 1998;**26**:31-34

[49] Small P, Blankenhorn D, Welty D, Zinser E, Slonczewski JL. Acid and base resistance in Escherichia coli and Shigella flexneri: Role of rpoS and growth in acid. Journal of Bacteriology. 1994;**176**:1729-1737

[50] Price SB, Wright JC, DeGraves FJ, Castanie-Cornet MP, Foster JW. Acid resistance systems required for survival of Escherichia coli O157:H7 in the bovine gastrointestinal tract and in apple cider are different. Applied and Environmental Microbiology. 2004;**70**:4792-4799

[51] Richard H, Foster JW. Escherichia coli glutamate- and arginine-dependent acid resistance systems increase internal pH and reverse transmembrane potential. Journal of Bacteriology. 2004;**186**:6032-6041

[52] Kühle AV, Jespersen L, Glover RLK, Diawara B, Jakobsen M. Identification and characterization of Saccharomyces cerevisiae strains isolated from West African sorghum beer. Yeast. 2001;**18**:1069-1079

[53] Naumova ES, Korshunova IV, Jespersen L, Naumova GI. Yeast research. Federation of Europe Microbiological Societies. 2003;**3**:177-184

[54] N'Da AK, Coulibaly S. Etude comparative de procédés traditionnels de préparation d'une bière locale appelée tchapalo. 2ème journées annuelles de la SOACHIM, Dakar-Sénégal. 1996: p. 40-46

[55] Dombek KM, Ingram LO. Ethanol production during batch fermentation with saccharomyces cerevisiae : Changes in glycolytic Enzymes and internal pH. Applied and Environmental Microbiology. 1987;**53**(6):1286-1291

Chapter 9

Electronic Noses Applications in Beer Technology

José Pedro Santos, Jesús Lozano and
Manuel Aleixandre

Additional information is available at the end of the chapter

http://dx.doi.org/10.5772/intechopen.68822

Abstract

This chapter describes and explains in detail the electronic noses (e-noses) as devices composed of an array of sensors that measure chemical volatile compounds and apply classification or regression algorithms. Then, it reviews the most significant applications of such devices in beer technology, with examples about defect detection, hop classification, or beer classification, among others. After the review, the chapter illustrates two applications from the authors, one about beer classification and another about beer defect detection. Finally, after a comparison with other analytical techniques, the chapter ends with a summary, conclusions, and the compelling future of the e-noses applied to beer technology.

Keywords: beer, electronic noses, gas sensor, pattern recognition, beer discrimination, defect detection

1. Introduction

Among the alcoholic beverages, beer is one of the most consumed in the world [1]. Because of such a big market, beer safety and the repeatability of its organoleptic qualities are very important for the manufacturers. For this reason, several techniques are used in the beer industry; they assess these aspects by analyzing the chemical composition or by human panels. One of the most difficult analyses is the chemical species of the beer headspace (HS). The headspace contains multitude of volatile organic compounds (VOC) that determine the organoleptic qualities or VOCs that are markers of spoilage, toxins, or flavor instability. For example, beer evolution can be marked by the presence of oxidative species such as aldehydes produced from alcohol oxidation [2]; contamination by bacteria can be marked by the presence of diketones such as diacetyl [3]; or toxic species such as nitrosamines [4] that can be generated during the malt production. Furthermore, the analysis is hindered by the high amount of CO_2 in the beer so usually, there is a required degasification process. The analysis of the volatile chemical species not only applies to the final product but also to the ingredients (hops, malt, etc) as the quality and characteristics of the components have a determinant effect on the final product. All these analyses (human and chemical) are complex so they are done in batches and are costly in time and money. However, an early detection or continuous monitoring by means of a fast analysis can avoid wastage and save money and time. An e-nose is a technology that could able to analyze and extract the chemical information of the beer organic volatile compounds and correlate their organoleptic qualities. The aim of this chapter is to bring in the innovative and promising technology of e-noses in the traditional beer industry. With this intention, there is, first, an introduction to e-nose technology where the different parts are described in detail. Then, several examples of its use are enumerated and some of them are explained. In addition, there is a brief comparison with other techniques to highlight the advantages and the scope of its use. Finally, there are the conclusions and future trends.

2. Electronic noses

An e-nose is a device that tries to imitate the structure and functionality of the human nose, and both of them are intended to be used in similar applications. **Figure 1** shows this similarity in the way they work: the first step in both is the interaction between volatile compounds (usually a complex mixture) with the appropriate receptors: olfactory receptors in the biological nose and a sensor array in the case of the e-nose. There is an overlapping in sensitivities so one odorant receptor responds to multiple odorants and one odorant can be detected by

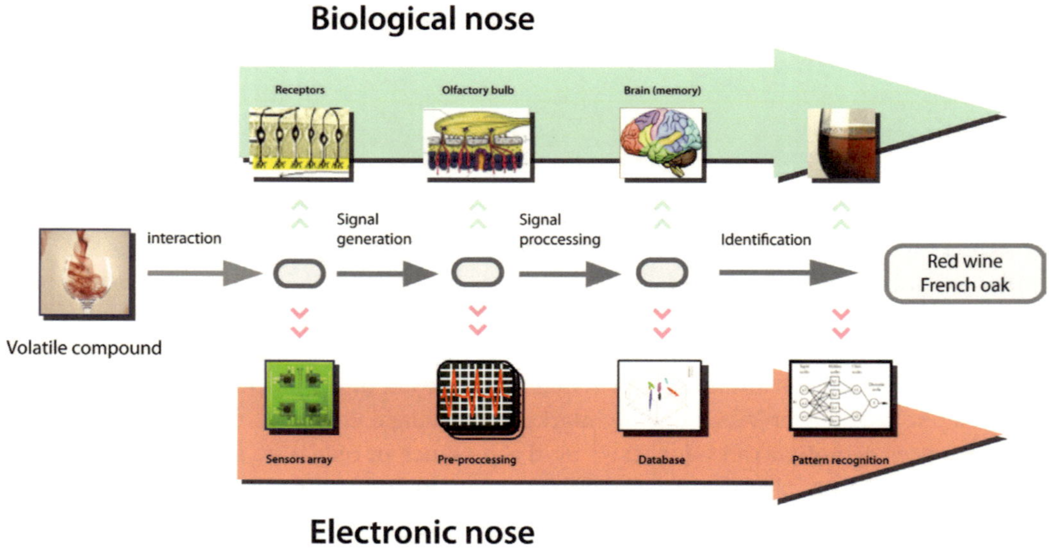

Figure 1. Similarity between the biological olfactory system and an e-nose.

several odorant receptors. The next step in both is the storage of these signals in the brain or in a database of a pattern recognition machine (learning stage) for further identification of one of the odors that is learned (classification stage).

Pre-1920 work on machine olfaction was prevented by the absence of any suitable electronic instrumentation. In 1920, Zwaardemaker and Hogewind suggested that odors could be detected by measuring the electrical charge developed on a fine spray of water that contained the odorant in solution, but they were unable to develop this into a useful instrument [5].

The first real report of an experimental instrument was published by Hartman and colleagues who described an electrochemical sensor consisting of a polished metal wire microelectrode in contact with the surface of a porous rod saturated with a dilute electrolyte [6, 7]. By using various combinations of metal electrodes, electrolytes, and applied potentials, a system of several sensors was made to operate simultaneously. In essence, the sensors used in this work were examples of amperometric electrochemical gas sensors. The instrument they developed comprised an array of eight different electrochemical cells and gave different patterns of response for different odorant samples, although in this work, they made no serious attempt to process the patterns which they generated even though computers were becoming available.

At about the same time, Moncrieff was working on the same problem but using a different approach. He employed a single thermistor (temperature-sensitive resistor) coated with a number of different materials, including poly(vinyl chloride), gelatin, and vegetable fat, to monitor odors [8]. He recognized that the coatings he used were non-specific, and he postulated that if an array of six thermistors with six different coatings were constructed, then the resulting instrument would be able to discriminate between a large number of different smells. In 1965, two other groups published early studies of e-noses: Buck et al. made use of the modulation of conductivity [9], while Drawnieks and Trotter used the modulation of the contact potential to monitor odors [10]. However, the concept of an e-nose as an intelligent system composed of an array of chemical sensors for odor identification did not emerge until nearly 20 years later, following publications by Persaud and Dodd in 1982 at Warwick University, UK [11] and by Ikegami in 1985 and 1987 at the Hitachi Research Laboratory in Japan [12, 13]. By this time, developments in electronics, sensors, and computing came together to reach a stage where an e-nose had become a genuine possibility. The term "e-nose" first appeared in the literature around the late 1980s [14]. Then, in 1991, a session of a NATO advanced workshop on chemosensory information processing was dedicated to the topic of artificial olfaction.

An accepted definition of an e-nose given by Gardner in 1994 is "an instrument which comprises an array of electronic chemical sensors with partial specificity and an appropriate pattern recognition system, capable of recognizing simple or complex odours" [15]. This definition restricts the term e-nose to those types of sensor array systems that are specifically used to sense odorous molecules in an analogous manner to the human nose. However, the architecture of an e-nose has much in common with multi-sensor systems designed for the detection and quantification of individual components in a simple gas or vapor mixture [16]. An e-nose generally consists of an aroma extraction system, a sensor array, a control and measurement system, and a pattern recognition method [17]. A block diagram of the typical structure of an e-nose can be observed in **Figure 2**.

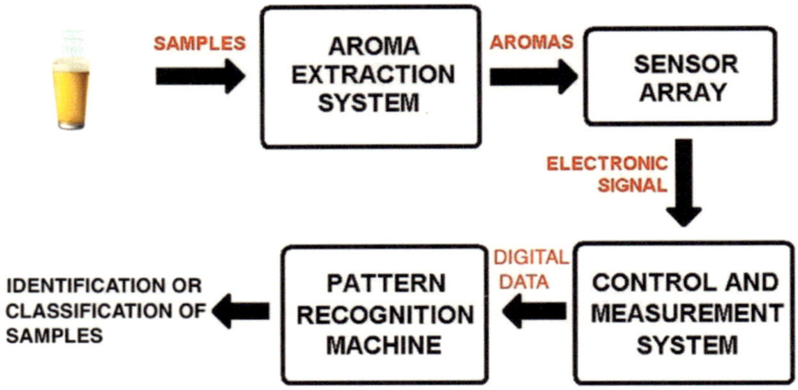

Figure 2. Block diagram of an e-nose system.

The sampling method or aroma extraction system carries the aromatic volatile compounds from the samples to the sensor cell. Different sampling techniques can be used in e-noses [18]: static or dynamic HS, purge and trap (P&T), and solid-phase micro-extraction (SPME) are the most common techniques.

A gas sensor is a device that is capable of converting the concentration of chemical compounds into electric signals and responds to the concentration of specific particles in gases or liquids [19]. Chemical sensors can be based on electrical, thermal, mass, or optical transducer principles. Several examples of chemical sensors used in e-noses are conducting polymers [20], quartz resonators [21], and surface acoustic wave (SAW) [22] and semiconductor devices [23]. The e-nose device has the advantage of low cost and portability for making in situ and online measurements.

The instrumentation and control system includes the electronic circuits needed for the measurements of sensors signals (e.g., interface circuits, signal conditioning, and analog-to-digital (A/D) or digital-to-analog (D/A) converters).

The goal of an e-nose is to identify an odorant sample and perhaps to estimate its concentration by means of a signal processing and pattern recognition system. However, those two steps may be subdivided into the following steps [24]: preprocessing, feature extraction, prediction or classification, and decision-making.

It is necessary to create a database of expected odorants by presenting the samples to the sensors. Preprocessing techniques try to compensate sensor drift, compress the transient response of the sensors, and reduce variations from sample to sample. Typical techniques include manipulation of baselines, response normalization, and compression of sensor transients.

Feature extraction has two different purposes: first, to reduce the dimensionality of the measurement space and second, to extract information relevant for pattern recognition. It is generally performed with linear transformations such as the classical principal component analysis (PCA) and linear discriminant analysis (LDA). PCA is a powerful, linear, unsupervised, and non-parametric pattern recognition technique that has been used by many researchers to reduce the dimensionality of the pattern space leading to better visualization of data clustering. This method expresses the response vectors by using linear combinations of orthogonal vectors along a new set of axes and is sometimes referred to as vector decomposition, and it usually helps to display multivariate data in a plot of two or three dimensions. PCA finds projection of maximum variance and is the most widely used linear feature-extraction technique [24]. But it is not optimized for classification tasks since it ignores the identity (class label) of the odor examples in the database. LDA, on the contrary, tries to find projections that maximize the distance between examples from different odorants and minimize the distance between examples of the same class [25, 26].

Finally, the classification task is usually performed by artificial neural networks (ANNs). An ANN is a mathematical algorithm that has the same function as that of the human brain in the biological sense of smell. The typical structure of an ANN is a network with two or more layers of neurons that are connected with synaptic weights—real number multipliers that connect the output of neurons to the inputs of neurons in the next layer [25–27]. During training, the ANN tries to learn the patterns of the different odorants by adapting the weights in order to obtain the desired output. After training, when an unidentified sample is presented, the ANN calculates the output of each layer and assigns the class label that provides the best response. In some cases, an undetermined class is used to determine the unknown sample that does not belong to any of the learned classes in the database [28].

The application of the human sense of smell as an odorant instrument is limited for different reasons: it is strongly subjective, gets tired easily, and in some cases, it is difficult to interpret. Consequently, there is considerable need for an instrument that could mimic the human sense of smell but without those limitations in order to be used as industrial applications. In this sense, e-noses could be used in areas like food, automobile, environmental industry, and medicine for different tasks like: pollution control and air-quality monitoring, control of industrial processes, detection of illnesses by exhaled breath, and safety aspects.

E-nose instruments are attractive in different fields due to several reasons: the fast assessment of samples, a qualitative and quantitative representation, and the use of low-cost and small-size sensors, appropriate for production processes. A considerable number of applications of e-noses have been reported for sensing applications. A broad list of e-nose reviews can be found in the literature that are structured and focused on mass spectrometry-based e-noses [29], biomedical and health care applications [30], agriculture and forestry applications [31], microbial quality control of food products [32] and food industry [33], pharmaceutical applications [34] developing chemical sensor arrays [35], etc.

3. E-nose applications in beer

E-noses have been applied to the whole production chain of beer, from the main ingredients (barley, hops, and yeasts) going through the different processes (mainly fermentation) to the final product. In this sense, the discrimination between beers and other beverages is the one that has more applications with enoses. In the following lines, an overview of the different applications of e-noses in breweries is given although it does not pretend to be exhaustive. A brief review of the applications of e-nose technology in breweries can be found in Ref. [36].

A classification of beer ingredients has been studied with these systems. Hops classification was performed in Ref. [37]. Using a commercial e-nose (FOX 2000), discrimination of different malt types could be done [38].

Fermentation monitoring, usually by measuring alcohol content, has been addressed by some authors [39, 40]. A recent review of e-nose applications on alcoholic beverage fermentation is found in Ref. [40].

Other common applications are detection of defects such as dimethyl sulfide detection [41] and 1-hexanol, ethyl acetate, oct-1-en-3-ol and diacetyl [42]. Related to this is the work in Ref. [43] that studies the influence of the fungicide triadimefon on the quality of beer using an e-nose, e-tongue, high-performance liquid chromatography (HPLC), and a sensory panel. Other authors [44] used a hand-held e-nose to detect different amounts of intentionally added ethyl acetate and acetaldehyde to commercial beer. In Ref. [45], discrimination among beers with some defects (diacetyl or dimethylsulfide) was performed using an MOS-based e-nose.

But by far the most common application is beer discrimination among other beers or among other alcoholic beverages. The first work about e-nose applied to beer is the paper by Aishima [46]. He used an e-nose composed of six Taguchi metal-oxide (MOX) gas sensors to discriminate among several alcoholic beverages. Also, in the 1990s, an e-nose based on conducting polymers was used to classify different types of beers and identify certain off-flavors [47]. In Ref. [48], an e-nose based on quartz crystal balance (QCB) and MOSX sensors was used in conjunction with a chromatographic column to discriminate among eight brands of beer. Other authors used an MOX-based e-nose to classify several alcoholic beverages and studied the ethanol influence in the response to the system [49]. In Ref. [50], an MOX-based e-nose was also used to differentiate among beverages, including beer, and in Refs. [51, 52], an MOX-based e-nose was used to classify commercial beers. Li [53] used an MS-based e-nose to discriminate among beers with different characteristics. Zhou discriminated among 12 commercial beers and other products with a 3 MOX sensor-based e-nose [54]. Siadat used an MOX-based e-nose to differentiate among alcoholic and non-alcoholic beer [55]. Vera [56] used an MS e-nose to discriminate among several beers from different breweries. Recently, Liu used a combined e-nose and electronic tongue system to correlate the information with a sensory panel for five commercial beers [57]. A SAW-based e-nose to discriminate volatile compounds, liquors, and perfumes can be found in [58].

Beer-aging experiments in commercial (lager and ale beers) industries under different storage conditions are shown in Refs. [59, 63, 64]. A review of e-noses with some information for beer classification and ageing is shown in Ref. [60].

Another application, far from the above cited, is the example of Thepudom et al. [61], who employed an optical e-nose to investigate alcohol decay in breath after drinking beer.

Table 1 summarizes the applications of e-noses in beer with information about the technology, number of sensors, and data processing algorithms.

Application	Sensor technology	Number of sensors	Data processing algorithm	References
Malt classification	MOX	6	PCA, CLA	[38]
Hops classification	MOX	6	PCA, SOM	[37]
Classification	CP	12	CLA	[47]
	SAW	8	PCA	[58]
	GC-MOS	14	PCA	[49]
	MOX	3	ANN	[50]
	MOX	8	PCA	[46]
	QCB, MOX	8,8	PCA	[48]
	MOX	5	PCA, LDA, PNN,	[51]
	MS	1	PLS-DA, SIMCA	[53]
	MOX	5	PCA	[52]
	MOX	3	PCA	[54]
	MOX	5	PCA, ANN	[55]
	MS	1	PCA, ANN	[56]
			PCA, LDA	
Off defects	CP	?	ANN	[45]
	MS	1	?	[41]
	GC-MOS	14	PCA, DFA	[42]
	MOS	4	ANN	[44]
Fermentation	Catalytic	1	?	[39]
	MOX	7	PCA	[62]
Flavor assessment	MOS	8	ANN	[57]
Aging	MS	1	PCA, LDA, PLS	[59]
	MOX	12	PCA	[63]
	MOX	5	PCA, LDA, ANN	[64]
Other (pesticides)	MOX	10	PCA	[43]
Breath	OPT	2	PCA	[61]

Table 1. E-noses and their beer applications.

4. Detailed examples of applications in the beer field

In this part of the chapter, the development of a portable e-nose designed and optimized for beer discrimination is presented. The device is validated by doing two different measurements: draft beer discrimination and off-odor detection. The main features of the designed e-nose [65] are portable, low size, autonomous, low cost, and easy to use. The designed e-nose (**Figure 3a**) is equipped with wireless communication capable of forming a network of e-noses for distributed measurements [66] (**Figure 3b**). It has been designed to work with resistive sensors, headspace as the sampling method, and a portable instrumentation and control system; it includes rechargeable batteries, touch screen, and IEEE 802.11 transceiver for wireless communication (**Figure 4**).

It consists of two gas inlets that are switched through a three-way electrovalve whose output is connected to the sensors cell that contains the micro-sensor array. One of the gas inlets has a carbon filter and is intended to provide clean air as the reference baseline. Downstream are located

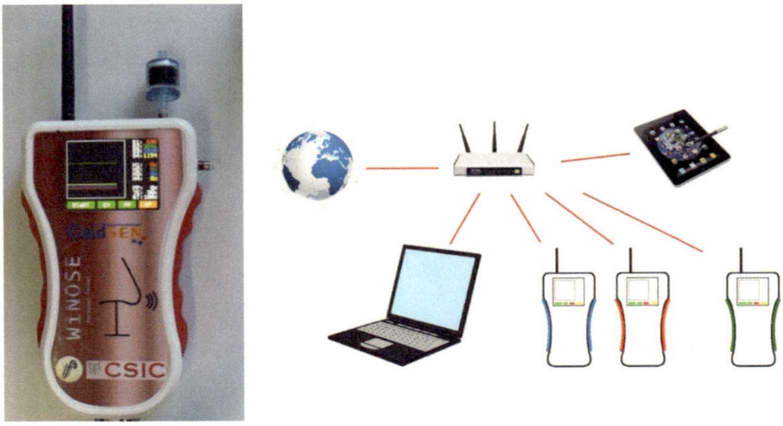

Figure 3. Portable e-nose (left) and configuration of the e-nose in the network (right).

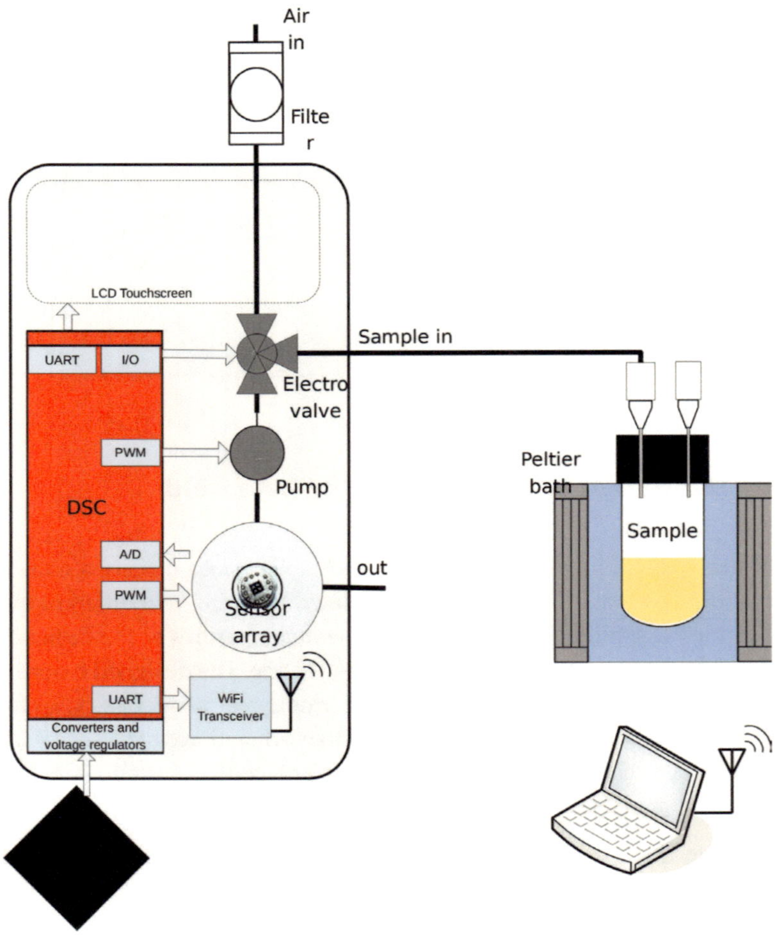

Figure 4. Schematics of the e-nose and measurement system.

the temperature sensor and the pump. The whole system is controlled by a digital signal controller (**Figure 5**). The sensor resistances are measured by A/D circuits, and their heating consists of pulse width modulation (PWM) outputs. An LCD touchscreen shows the measurements that allow the manual control of parameters such as pump, heater, power, or the electrovalve. Rechargeable batteries give about 8 h of autonomy to the e-nose. Wireless communications are provided by a Wifi transceiver. A network could be established with a host computer for remote operation. The sensors cell and board are designed for micro-sensors in a TO-5 or TO-8 12 leads package but it is easily adaptable to other packages and sensors. The system can measure up to four resistive sensors and provides independent heating for each one. The instrument is controlled by a program developed in Labview™. The program displays and controls the measurement parameters and generates the response database. Algorithms for both online and off-line pattern recognition techniques have been developed in Matlab™ and integrated in the program through Mathscripts. External classification using a web server can also be performed [67].

The control program displays and controls the main measurement parameters (temperature, sensor resistance, temperature, valve status, battery status, and pump power) and automatically generates the response database. The program user interface is shown in **Figure 6**.

After measurements are made, data processing methods are applied to the data. PCA and ANN have been implemented in Matlab™ for data processing. PCA applies a linear transformation to the data, and this results in a new space of variables called principal components [24]. The number of variables is reduced from the number of sensors 4 to 2 or 3 variables in order to show it in a plot and see the discrimination capability of the array. Next, a classifier is used to give a response to a typical problem of prediction of unknown samples. The most used classifiers are based in ANN. In these experiments, two types of ANNs were used for

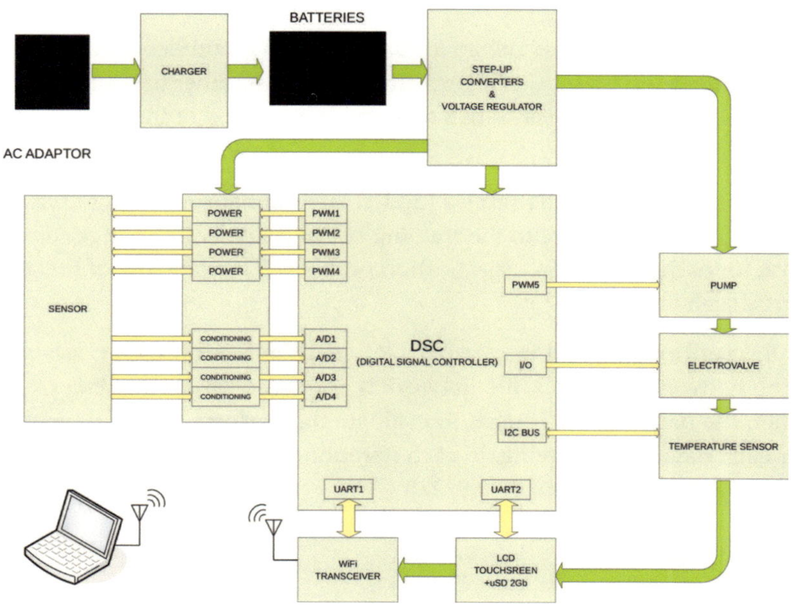

Figure 5. Block diagram of the main components of the developed portable e-nose.

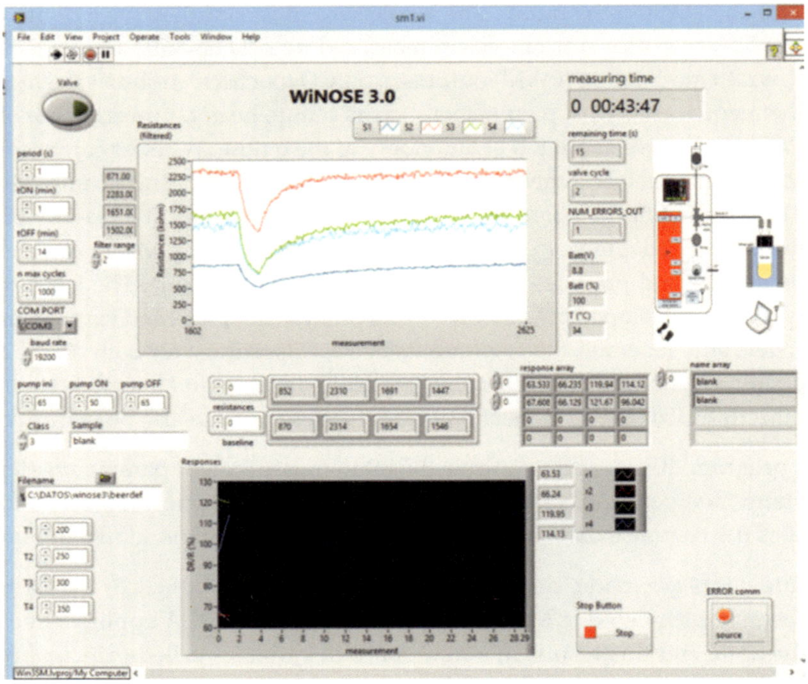

Figure 6. Control program user interface.

classification purposes: feedforward network with backpropagation learning algorithm (FF-BP) and probabilistic neural networks (PNN).

The FF-BP networks had three layers, an input layer with a number of neurons equal to the sensors of the e-nose (four), a hidden layer with a variable number of neurons that went from 10 to 25, and finally an output layer with a neuron for each class of the classification problem.

The PNN were also composed of three layers [26], the input layer with a number of neurons equal to the sensors of the e-nose (four), a hidden layer of neurons with radial basis transfer functions that had neurons equal to the training set, and finally an output competitive layer with a neuron for each class of the classification problem. The structure of this type of neural network can be seen in **Figure 7**.

To validate the performance of the network [26], leave one out (LOO) cross-validation was applied to the networks. In LOO, the network is trained with all the data except one data point and then the data left out is used to evaluate the performance of the network. This is repeated for each data point, leaving, in each iteration, one data point out. The performance is the assembled errors that are made [68, 27].

4.1. Beer discrimination

The main aim of this experiment was to check the discrimination capability of the proposed system. In this sense, the task of classifying different draft beers was attempted. Different

http://dx.doi.org/10.5772/intechopen.68822

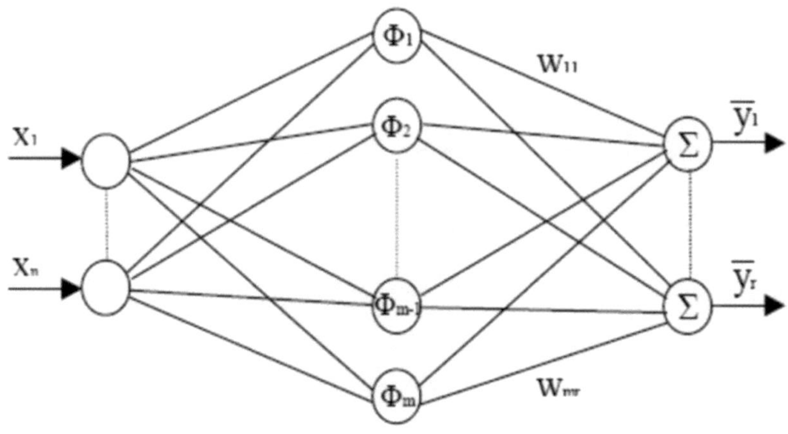

Figure 7. Scheme of the probabilistic neural network classifier.

commercial draft beers (Blomberg Blanca, Blomberg Rubia, Blomberg Dubbel, Marwan, Cerex, Jacha Jigo Jiguera, Ballut Rubia, and Ballut Negra) were purchased in specialized shops and kept to the moment of measurements. Before that, a degasification process based on magnetic agitation for 20 min was performed. Next, 10 mL of each sample was taken and kept at 12°C with a thermal bath in order to generate a stable headspace for each sample. A total number of 128 measurements (16 replicates per beer) were performed. The measurement cycle was for a duration of 10 min: 1 min of adsorption (air passes through the samples) and 9 min of desorption (air directly passes through the sensors). The air rate flow was 150 mL/min, constant in every measurement. Once a second, measurements of the parameters (sensors resistance, relative humidity, ambient temperature, air rate flow, and battery voltage) were taken and stored in a file. Commercial sensors of e2V SGX sensortech based on tin oxide MOS sensors were used. The operation temperatures of the sensors were optimized for beer discrimination, and its range was among 350 and 450°C. Once the measurements were performed, data were stored in a hard disk for data processing. A periodic calibration made with 10% of Ethanol in water is usually performed for compensating sensor drift. In this case, it is not necessary because of the short time spent doing the measurements. No variation in the sensors response to reference air is observed during the measurement period.

Data obtained from measurements were processed using PCA. The first three principal components were shown in **Figure 8**. This plot shows an almost complete separation among the eight classes of beer. The ellipses show 80% of the variance of the classes, some partial overlapping between Marwan and Cerex, and other neighbor classes can be observed. The variance explained by each principal component is in brackets.

To confirm these results, a classification with three different classifiers based on artificial intelligence was performed. Three different classifiers were used: feedforward neural network with backpropagation algorithm, probabilistic neural networks, and fuzzy logic (FL) classifiers. Both FF-BP and PNN employed eight neurons at the output layer corresponding with the eight brands of craft beers. In the case of the fuzzy logic-based classifier, a total of eight

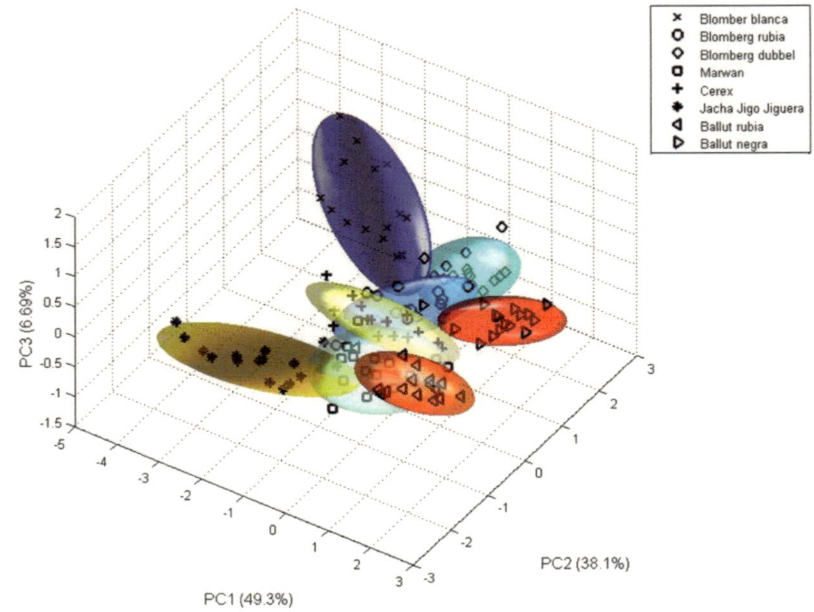

Figure 8. PCA score plot of the measurements of different draft beers.

fuzzy rules (corresponding with the eight brands of beers) is built over each sensor output on the 128-sample knowledge database. Each of these rules is optimized in the training stage to maximize both acceptance and rejection scores of the unknown samples. First-order cross-validation (leave one out) was used for validation because a great number of measurements were not available.

The confusion matrix obtained in the validation of the classifiers is shown in **Table 2**, in which the system classifies (in columns) the real samples (rows). The success rate obtained in the classification, defined as the rate between the number of samples correctly classified over the total number of measurements, was 87.5% for FF-BP network, 92.96% for the PNN, and 89.84% for FL-based classifier. Results confirm that the PNN classifier presents the best performance in the classification of these beer samples.

4.2. Beer defects detection

Another experiment was made with the same prototype. In this case, two aromatic defects in beer (acetaldehyde and ethyl acetate), at a level between the organoleptic threshold and five times this quantity, were measured. Acetaldehyde threshold is 25 ppm and ethyl acetate is 21 ppm [69, 70]. The lager beer from cans was magnetically stirred (350 rpm, 20 min) to degas before the measurement procedure. Glass vials of 22 mL were filled with 10 mL of the sample and they were kept at 18 ± 1°C. A minimum of 10 replicates for each compound were measured.

Predicted real	Blomberg Blanca			Blomberg Rubia			Blomberg Dubbel			Marwan			Cerex			Jacha Jigo Jiguera			Ballut Rubia			Ballut Negra		
Blomberg Blanca	16	16	15						1															
Blomberg Rubia				14	15	13			1				1	1	2		1					1		
Blomberg Dubbel			1				16	15	15														1	
Marwan										15	11	13	1	6	2						1			
Cerex					4	1				2			14	12	14			1						
Jacha Jigo Jiguera										2	2	1				14	14	15						
Ballut Rubia										1	1	1							15	15	15			
Ballut Negra					1							1	1	2								15	14	15

Table 2. Confusion matrix with three classifiers: FF-BP, PNN, and FL.

Once the measurements have been made and the raw data is stored, data processing is performed. **Figures 9** and **10** show the 3D plot of a PCA made to the measurements for the ethyl acetate and acetaldehyde samples. The blank samples are clearly separated from the samples with the two compounds but the two defective types of samples (with the different compounds) seem to overlap a bit.

Results are confirmed with a non-linear classification method. PNN networks, which obtained the best performance in previous experiments, were trained to classify samples according to their defects. The percentages of cases correctly classified in the LOO validation were 83 and 91% for ethyl acetate and acetaldehyde, respectively. While all the samples at five times the threshold concentration values were correctly classified, there were some errors in the lower concentration samples; for ethyl acetate, the network confounded some blank samples with concentration of 1 T and also concentrations of 1 T with concentrations of 2 T; for acetaldehyde, only 2 samples of T concentrations and 5 T concentrations were confounded by the PNN.

A qualitative classification of the beers according to the defect (at all the concentrations levels) can be seen in **Figure 11**. The PCA shows that the beer samples with ethyl acetate, the beer samples with acetaldehyde, and the beer samples without defects (blank) are separated with only a small overlap among the classes.

For this qualitative classification problem, the PNN analysis gave a 94% success rate in the validation, regardless of the overlap seen in the PCA plot.

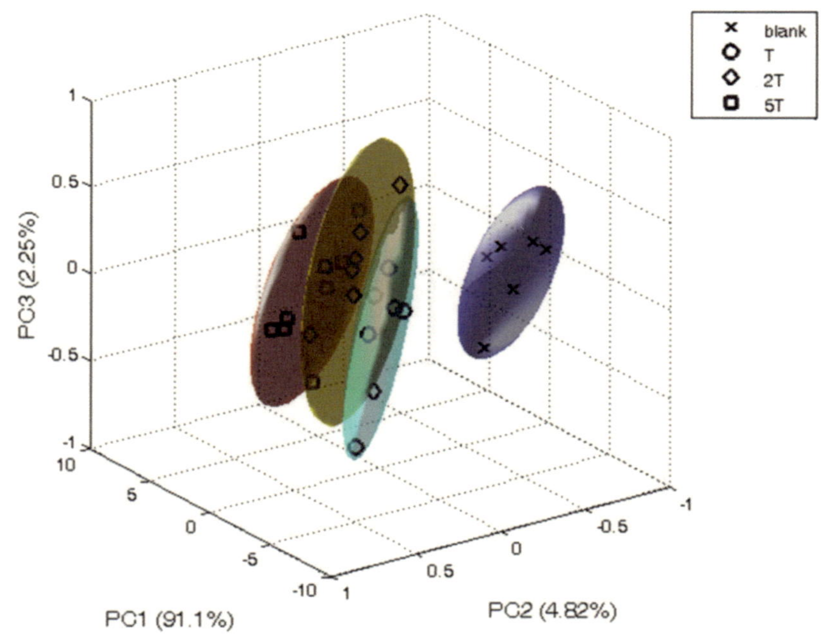

Figure 9. PCA plot for the ethyl acetate measurements in beer.

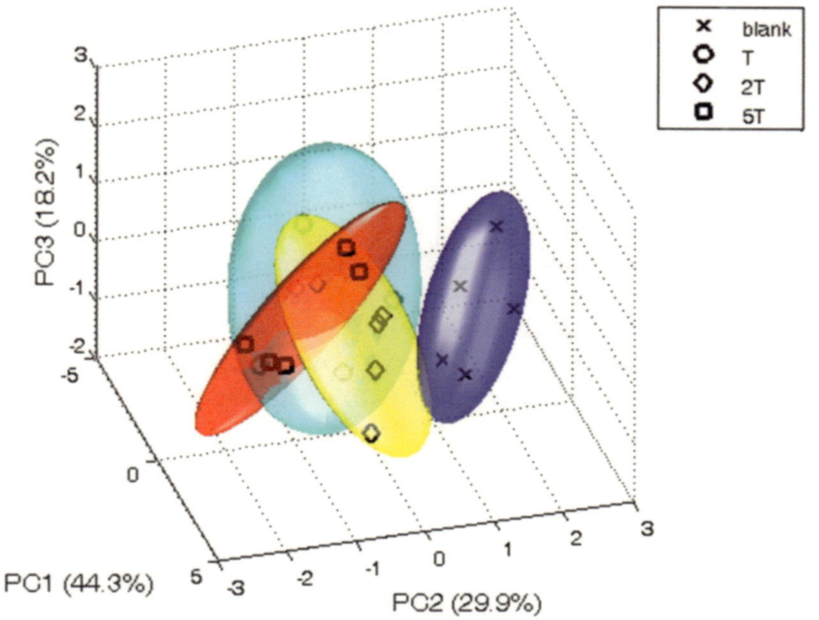

Figure 10. PCA plot for the acetaldehyde measurements in beer.

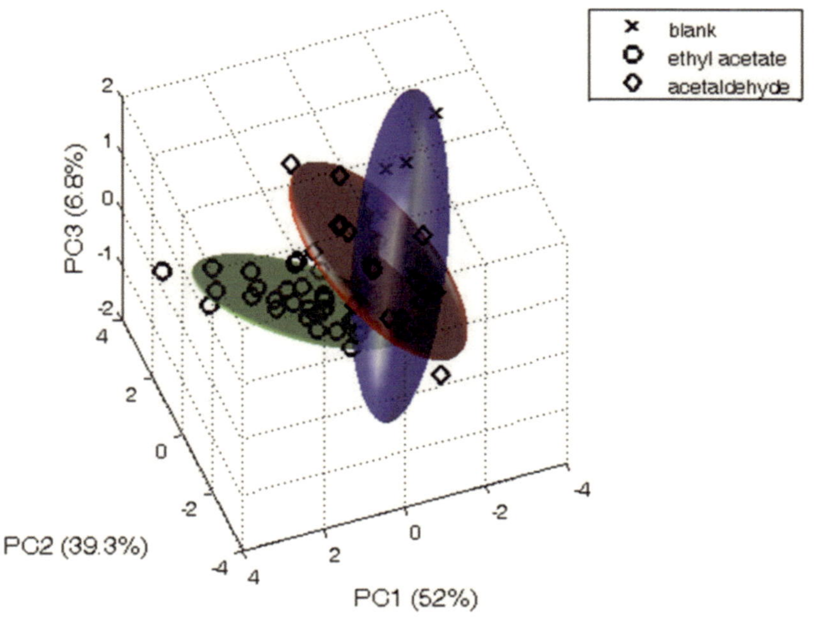

Figure 11. PCA plot for the defect measurements in beer.

These classification results could be improved by several strategies. Better classification algorithms could be applied to the data. Also, the data could be less noisy by using better systems that keep temperature or flows more controlled.

5. Comparison with other techniques

There are several traditional chemical methods like titration, gravimetric analysis, international bitterness units (IBUs), alcohol measurement, extract, calories, pH, high-performance liquid chromatography, or new methods like capillary electrophoretic method [70] to characterize the beer but few methods are used to analyze the volatile compounds of its headspace as the e-nose does. Techniques such as gas chromatography (GC), soft ionization techniques, or human sensory panels are among them. Each one has its own characteristics and when deciding what method to use, one should select the one more appropriate for its interest and the task at hand.

Human panels record the experience of beer tasting by several senses, and they use the most complex of them, the scent, to detect the volatile organic compounds that emanate from beer [71]. However, human panels are subject to variability not only within different panels but also with itself over time. They need to be trained to become experts which is time-consuming and expensive. Due to VOCs matrix interaction and physiological singularities, the correlation between the chemical analysis and the scents is not clear and not always straightforward [72] so human panelists are usually mandatory.

Traditional gas chromatography has been widely used for beer analysis, detecting singular VOCs. There are very different ways to gather, separate, and detect the volatile compounds [73]. GC has been used successfully to detect diacetyl, pentanedione, acetoin, and acetaldehyde during the fermentation process, showing the evolution of these VOCs [74]. GC monitors the concentration of singular compounds and requires time-consuming analysis to extend the analysis to the many volatile compounds that the beer headspace has. The process is slower, more expensive, and more complex, making it very difficult to operate in a continuous way.

GC-olfactometry combines both the human panel and gas chromatography by placing a human panel as a detector at the end of the chromatograph. In this way, the system can detect individual aromas and correlate them with the physiological sensation they provide. In [75], the compounds from Challenger and Saaz hops were analyzed and correlated with several olfactory descriptors in pellets and they analyzed the evolution after the brewing process. This technique allows the evaluation of the scent as perceived by humans but the analysis is costly and time-consuming.

Soft ionization techniques, selected ion flow tube mass spectrometry (SIFT-MS), or proton transfer reaction time-of-flight (PTR-ToF) allow fast analysis of the samples in a continuous fashion. For example, PTR-ToF has been used to monitor the fermentation process of different yeasts pointing to the different VOCs released [76], and SIFT-MS has been used to determine the aldehyde content of malt as a biomarker to identify each variety [77]. PTR-ToF usually lacks on precision in identifying the individual VOCs, and SIFT-MS has lower sensitivity.

Both methods are fast and can be used as online monitoring devices but they are cumbersome and expensive compared to the potential of e-noses.

6. Conclusions and future trends

Monitoring or fast analysis of the chemical composition of the beers in all steps of the production and consumption chain is of capital importance for the beer industry. Rapid detection of problems like contamination, spoiled ingredients, or flavor instability can save time, money, or prevent health problems. The traditional chemical beer analyses are done in batches; in contrast, the e-noses are a good instrument to use in online analyses. Although e-noses are not as accurate as traditional analysis, they work very well in controlled environments like the ones of the beer industry or in laboratory. In these controlled environments, differential analysis can be easily made and interesting results can be obtained like the ones presented in this chapter.

For more complex applications or scenarios, there is still a need for better sensors with better characteristics such as reproducibility, repeatability, and selectivity. There is also a need for the standardization of these sensors and methods to increase reproducibility of analysis so studies can easily be transferred from one device to another. But the sensor technology is blooming and multitude of research groups and enterprises are working on their improvement. New materials like carbon nanotubes, nanostructured metal oxides, or graphene materials are offering very interesting and improved sensor capabilities. To complement the sensor, new algorithms are being developed in deep learning or semi-supervised learning that would reach soon the e-nose technology and take advantage of those improved sensor characteristics. New commercial sensors are being miniaturized more and more and will be able to be deployed in small power consumption devices as wearables or mobile devices that will integrate in the Internet of things.

There are a multitude of possible applications, from the analysis of the ingredients to mobile applications for the final consumer. To name a few, there are several interesting possibilities that could be used in the analysis of ingredients, to detect spoilage, or to store scent profiles year after year. They could be used for the online monitoring of the fermentation process and by applying closed loop control to it. They could be used as storage monitoring, ensuring that the product reaches the final consumer in optimal conditions. The final consumer could use these new small sensors in wearable devices and apply them in food safety, detecting contaminants before consuming them. The e-nose presents endless and very interesting potential applications.

Abbreviations

ANN	Artificial neural network
CA	Correlation analysis
CDA	Canonical discriminant analysis

CP	Conducting polymers
CLA	Cluster analysis
DA	Discriminant analysis
DFA	Discriminant factorial analysis
FL	Fuzzy logic
GA	Genetic algorithms
GC	Gas chromatography
LR	Linear regression
LDA	Linear discriminant analysis
MOX	Metal oxide semiconductor sensor
MS	Mass spectrometry
OPT	Optical projection tomography
PCA	Principal component analysis
PLS	Partial least squares
QCB	Quartz crystal balance
SAW	Surface acoustic wave
SIMCA	Soft independent modeling by class analogy
SOM	Self-organizing maps

Author details

José Pedro Santos[1]*, Jesús Lozano[2] and Manuel Aleixandre[1]

*Address all correspondence to: jp.santos@csic.es

1 GRIDSEN, Instituto de Tecnologías Físicas y de la Información, Consejo Superior de Investigaciones Científicas, Madrid, Spain

2 Escuela de Ingenierías Industriales, Universidad de Extremadura, Badajoz, Spain

References

[1] World Health Organisation. Global Status Report on Alcohol and Health 2014. Global Status Report Alcohol [Internet]. 2014. pp. 1-392. Available from: http://www.who.int/substance_abuse/publications/global_alcohol_report/msbgsruprofiles.pdf

[2] Vanderhaegen B, Neven H, Verachtert H, Derdelinckx G. The chemistry of beer aging—a critical review. Food Chemistry. 2006;**95**(3):357-381

[3] Wainwright T. Diacetyl—a review: Part I—analytical and biochemical considerations: Part II—brewing experience. Journal of the Institute of Brewing [Internet]. 1973;**79**(6):451-470. Available from: http://doi.wiley.com/10.1002/j.2050-0416.1973.tb03567.x. [Accessed: March 1, 2017]

[4] Izquierdo-Pulido M, Barbour JF, Scanlan RA. N-nitrosodimethylamine in Spanish beers. Food and Chemical Toxicology (Internet). 1996;**34**(3):297-299. Available from: http://linkinghub.elsevier.com/retrieve/pii/0278691595001166. [Accessed: March 1, 2017]

[5] Zwaardemaker H, Hogewind F. On spray-electricity and waterfall-electricity. Proceedings of the Royal Academy of Sciences at Amsterdam. 1920;**22**

[6] Hartman JD. A possible objective method for the rapid estimation of flavors in vegetables. Proceedings of the American Society for Horticultural Science. 1954;**64**

[7] Wilkens WF, Hartman JD. An electronic analogue for the olfactory process. Annals of the New York Academy of Sciences. 1964;**116**

[8] Moncrieff RW. An instrument for measuring and classifying odours. Journal of Applied Physiology. 1961;**16**

[9] Buck TM, Allan FG, Dalton M. Detection of chemical species by surface effects on metals and semiconductors. In: Bregman JI, Dravnieks A, editors. Surface Effects in Detection. London: Macmillan; 1965

[10] Dravnieks A, Trotter PJ. Polar vapour detection based on thermal modulation of contact potentials. Journal of Scientific Instruments. 1965;**42**

[11] Persaud KC, Dodd GH. Analysis of discrimination mechanisms of the mammalian olfactory system using a model nose. Nature. 1982;**299**

[12] Ikegami A, Kaneyasu M. Olfactory detection using integrated sensors. In: Proceedings of the 3rd International Conference on Solid-State Sensors and Actuators (Transducers 85). New York: IEEE Press; 1985. pp. 136-139

[13] Kaneyasu M, Ikegami A, Arima H, Iwanga S. Smell identification using a thick-film hybrid gas sensor. IEEE Transactions on Components, Packaging, and Manufacturing Technology, CHMT-10. 1987;**10**

[14] Gardner JW. Pattern recognition in the Warwick electronic nose. In: 8th International Congress of the European Chemoreception Research Organisation; Coventry, UK. 1988. p. 9

[15] Gardner JW. A brief history of electronic noses. Sensors and Actuators B: Chemical. 1994;**1818-19**

[16] Gardner JW, Bartlett PN. Electronic Noses: Principles and Applications. Vol. 233. New York: Oxford University Press; 1999

[17] Lozano-Rogado J. , New Technology in Sensing Odours: From Human to Artificial Noses. In Floriculture, Ornamental and Plant Biotechnology, vol. IV, Global Science Books, UK, pp. 152-161, 2006

[18] Lozano J, Santos JP, Gutiérrez J, Horrillo MC. Comparative study of sampling systems combined with gas sensors for wine discrimination. Sensors and Actuators B: Chemical. 2007;**126**(2)

[19] Pearce T C, Schiffman S S, Nagle H T & Gardner J W. Handbook of machine olfaction: electronic nose technology. 2006. John Wiley & Sons, Weinheim (Germany)

[20] Guadarrama A, Fernandez JA, Iniguez M, Souto J, De Saja JA. Array of conducting polymer sensors for the characterisation of wines. Analytica Chimica Acta. 2000;**411**(1)

[21] Sharma, P., Ghosh, A., Tudu, B., Sabhapondit, S., Baruah, B. D., Tamuly, P. & Bandyopadhyay, R. (2015). Monitoring the fermentation process of black tea using QCM sensor based electronic nose. Sensors and Actuators B: Chemical, 219, 146-157

[22] Lozano J, Fernández MJ, Fontecha JL, Aleixandre M, Santos JP, Sayago I, Horrillo MC. Wine classification with a zinc oxide SAW sensor array. Sensors and Actuators B: Chemical. 2006;**120**(1)

[23] Di Natale C, Macagnano A, Nardis S, Paolesse R, Falconi C, Proietti E, D'Amico A. Comparison and integration of arrays of quartz resonators and metal-oxide semiconductor chemoresistors in the quality evaluation of olive oils. Sensors and Actuators B: Chemical. 2001;**78**(1)

[24] Gutierrez-Osuna R. Pattern analysis for machine olfaction: A review. IEEE Sensors Journal. 2002;**2**(3)

[25] Waldemark, J. T., Roppel, T. A., Wilson, D. M., Dunman, K. L., Padgett, M. L., & Lindblad, T. (1999, March). Neural networks and PCA for determining region of interest in sensory data preprocessing. In Ninth Workshop on Virtual Intelligence/Dynamic Neural Networks: Neural Networks Fuzzy Systems, Evolutionary Systems and Virtual Re (pp. 396-405). International Society for Optics and Photonics

[26] Duda, R O, Hart, P E, & Stork, D. G. Pattern classification. 2nd. Edition. Wiley & Sons. New York. 2001

[27] Bishop CM. Neural Networks for Pattern Recognition. Oxford: Oxford University Press; 1999

[28] Nagle HT, Gutierrez-Osuna R, Schiffman SS. The how and why of electronic noses. IEEE Spectrum. 1998;**35**(9)

[29] Peris M, Escuder-Gilabert L. A 21st century technique for food control: Electronic noses. Analytica Chimica Acta. 2009;**638**

[30] Wilson AD, Baietto M. Advances in electronic-nose technologies developed for biomedical applications. Sensors (Basel). 2011;**11**

[31] Wilson AD. Diverse applications of electronic-nose technologies in agriculture and forestry. Sensors (Basel). 2013;**13**

[32] Falasconi M, Concina I, Gobb E, Sberveglieri V, Pulvirenti A, Sberveglieri G. Electronic nose for microbiological quality control of food products. International Journal of Electrochemical Science. 2012;**2012**

[33] Loutfi A, Coradeschi S, Mani GK, Shankar P, Rayappan JBB. Electronic noses for food quality: A review. Journal of Food Engineering. 2015;**144**:103-111

[34] Alam H, Saeed SH, Engg C. Electronic nose in food and health applications: A review. International Journal of Computing and Corporate Research. 2012;**2**

[35] James D, Scott SM, Ali Z, O'Hare WT. Chemical sensors for electronic nose systems. Microchimica Acta. 2005;**149**

[36] Ghasemi-Varnamkhasti M, Mohtasebi SS, Rodriguez-Mendez ML, Lozano J, Razavi SH, Ahmadi H. Potential application of electronic nose technology in brewery. Trends in Food Science & Technology. 2011;**22**(4)

[37] Lamagna A, Reich S, Rodriguez D, Scoccola NN. Performance of an e-nose in hops classification. Sensors and Actuators B: Chemical. 2004;**102**(2)

[38] Zimmermann D, Leclercq C. Electronic nose for monitoring the flavour of special malts. In: 2nd International Symposium on Olfaction and Electronic Nose.; Tolouse. 1995

[39] Austin GD, Russell I, Meiering AG, Subden RE. A gas-sensor-based on-line ethanol meter for Breweries. Journal of the American Society of Brewing Chemists. 1996;**54**(4)

[40] Buratti S, Benedetti S. Alcoholic Fermentation Using Electronic Nose and Electronic Tongue, In Electronic Noses and Tongues in Food Science (pp. 291-299). Academic Press, San Diego, 2016

[41] Kojima H, Araki S, Kaneda H, Takashio M. Application of a new electronic nose with fingerprint mass spectrometry to brewing. Journal of the American Society of Brewing Chemists. 2005;**63**(4)

[42] Ragazzo-Sanchez JA, Chalier P, Chevalier-Lucia D, Calderon-Santoyo M, Ghommidh C. Off-flavours detection in alcoholic beverages by electronic nose coupled to GC. Sensors and Actuators B: Chemical. 2009;**140**(1)

[43] Kong Z, Li M, An J, Chen J, Bao Y, Francis F, et al. The fungicide triadimefon affects beer flavor and composition by influencing Saccharomyces cerevisiae metabolism. Scientific Reports. 2016;**6**

[44] Santos, J. P., & Lozano, J. (2015, February). Real time detection of beer defects with a hand held electronic nose. In Electron Devices (CDE), 2015 10th Spanish Conference on (pp. 1-4). IEEE.

[45] Bailey TP, Hammond RV, and Persaud KC. Application for an electronic aroma detector in the analysis of beer and raw materials. Journal of the American Society of Brewing Chemists. 1995;**53**

[46] Aishima T. Discrimination of liquor aromas by pattern recognition analysis of responses from a gas sensor array. Analytica Chimica Acta (Internet). 1991;**243**:293-300. Available from: http://linkinghub.elsevier.com/retrieve/pii/S0003267000825738. [Accessed: February 20, 2017]

[47] Pearce TC, Gardner JW, Friel S, Bartlett PN, Blair N. Electronic nose for monitoring the flavour of beers. Analyst. 1993;**4**(118)

[48] Heberle I, Liebminger A, Weimar U, Göpel W. Optimised sensor arrays with chromatographic preparation: Caracterisation of alcoholic beverages. Sensors and Actuators B: Chemical. 2000;**68**(1):53-57

[49] Ragazzo-Sanchez JA, Chalier P, Chevalier D, Ghommidh C. Electronic nose discrimination of aroma compounds in alcoholised solutions. Sensors and Actuators B: Chemical. 2006;**114**(2)

[50] Tao, Z., Lei, W., & Teng, J. (2008, December). Pattern recognition of the universal electronic nose. In Intelligent Information Technology Application, 2008. IITA'08. Second International Symposium on (Vol. 3, pp. 249-253). IEEE.

[51] Pornpanomchai, C., & Suthamsmai, N. (2008, August). Beer classification by electronic nose. In Wavelet Analysis and Pattern Recognition, 2008. ICWAPR'08. International Conference on (Vol. 1, pp. 333-338). IEEE.

[52] Ghasemi-Varnamkhasti M, Mohtasebi SS, Siadat M, Ahmadi H, Razavi SH. From simple classification methods to machine learning for the binary discrimination of beers using electronic nose data. Engineering in Agriculture, Environment and Food. 2015;**8**(1)

[53] Li W, Pickard MD, Beta T. Evaluation of antioxidant activity and electronic taste and aroma properties of antho-beers from purple wheat grain. Journal of Agricultural and Food Chemistry. 2007;**55**(22)

[54] Zhou, J., Lei, W., Teng, J., & Tao, Z. (2010, July). Research on the recognition of chemical odour information based on probabilistic neural networks. In Environmental Science and Information Application Technology (ESIAT), 2010 International Conference on (Vol. 1, pp. 825-828). IEEE.

[55] Siadat, M., Losson, E., Ghasemi-Varnamkhasti, M., & Mohtasebi, S. S. (2014, November). Application of electronic nose to beer recognition using supervised artificial neural networks. In Control, Decision and Information Technologies (CoDIT), 2014 International Conference on (pp. 640-645). IEEE.

[56] Vera L, Aceña L, Guasch J, Boqué R, Mestres M, Busto O. Characterization and classification of the aroma of beer samples by means of an MS e-nose and chemometric tools. Analytical and Bioanalytical Chemistry. 2011;**399**(6)

[57] Liu J, Yang J, Wang W, Fu S, Shi Y, Men H. Automatic evaluation of sensory information for beer at a fuzzy level using electronic tongue and electronic nose. Sensors and Materials. 2016;**28**(7)

[58] Yang YM, Yang PY, Wang XR. Electronic nose based on SAWS array and its odor identification capability. Sensors and Actuators B: Chemical. 2000;**66**(1):167-170

[59] Sikorska E, Chmielewski J, Górecki T, Khmelinskii I, Sikorski M, De Keukeleire D. Discrimination of beer flavours by analysis of volatiles using the mass spectrometer as an electronic nose. Journal of the Institute of Brewing. 2007;**113**(1)

[60] Berna A. Metal oxide sensors for electronic noses and their application to food analysis. Sensors (Basel) (Internet). 2010;**10**(4):3882-3910. Available from: http://www.ncbi.nlm.nih.gov/pubmed/22319332. [Accessed: February 13, 2017]

[61] Thepudom, T., Kerdcharoen, T., Tuantranont, A., & Pogfay, T. (2012, December). Healthcare electronic nose to detect beer odor in breath after drinking. In Biomedical Engineering International Conference (BMEiCON), 2012 (pp. 1-4). IEEE.

[62] Phetchakul T, Sutthinet C. Monitoring of draft beer fermentation process by electronic nose. Advanced Materials Research. 2014;**911**

[63] Mckellar RC, Young JC, Johnston A, Knight KP, Lu X, Buttenham S. Use of the electronic nose and gas chromatography-mass spectrometry to determine the optimum time for aging of beer. Technical Quarterly—Master Brewers Association Amsterdam [Internet]. 2002;**39**(2):99-105. Available from: http://cat.inist.fr/?aModele=afficheN&cpsidt=13842154. [Accessed: February 20, 2017]

[64] Ghasemi-Varnamkhasti M, Mohtasebi SS, Siadat M, Lozano J, Ahmadi H, Razavi SH, et al. Aging fingerprint characterization of beer using electronic nose. Sensors and Actuators B: Chemical. 2011;**159**(1)

[65] Lozano J, Santos JP, Suárez JI, Arroyo P, Herrero JL, Martín A. Detection of pollutants in water samples with a wireless hand-held E-nose. Procedia Engineering. 2014;**87**

[66] Santos JP, Lozano, J, Suárez JI, Aleixandre M. Electronic Nose Network For Detection Of TATP Precursors. Proceedings of International Simposium on Olfaction and Electronic Nose 2013. Daegu (South Korea). 2013

[67] Herrero JL, Lozano J, Santos JP, Suárez JI. On-line classification of pollutants in water using wireless portable electronic noses. Chemosphere. 2016;**152**:107-116

[68] Derks EP, Beckers ML, Melssen WJ, Buydens LM. Parallel processing of chemical information in a local area network—II. A parallel cross-validation procedure for artificial neural networks. Computers & Chemistry. 1996;**20**(4):439-448

[69] Blanco CA, Andrés-Iglesias C, Montero O. Low-alcohol beers: Flavor compounds, defects, and improvement strategies. Critical Reviews in Food Science and Nutrition. 2016;**56**(8)

[70] Cortacero-Ramírez S, de Castro MH-B, Segura-Carretero A, Cruces-Blanco C, Fernández-Gutiérrez A. Analysis of beer components by capillary electrophoretic methods. TrAC Trends in Analytical Chemistry (Internet). 2003;**22**(7):440-455. Available from: http://www.sciencedirect.com/science/article/pii/S0165993603007040

[71] Stucky GJ, Mcdaniel MR. Raw hop aroma qualities by trained panel free-choice profiling. Journal of the American Society of Brewing Chemists (Internet). **55**(2):65-72. Available from: http://cat.inist.fr/?aModele=afficheN&cpsidt=2677437. [Accessed: March 1, 2017]

[72] Lozano J, Santos JP, Aleixandre M, Sayago I, Gutiérrez J, Horrillo MC. Identification of typical wine aromas by means of an electronic nose. IEEE Sensors Journal. 2006;**6**(1)

[73] Andrés-Iglesias C, Montero O, Sancho D, Blanco CA. New trends in beer flavour compound analysis. Journal of the Science of Food and Agriculture (Internet). 2015;**95**(8):1571-1576. Available from: http://dx.doi.org/10.1002/jsfa.6905

[74] Tian J. Determination of several flavours in beer with headspace sampling–gas chromatography. Food Chemistry (Internet). 2010;**123**(4):1318-1321. Available from: http://www.sciencedirect.com/science/article/pii/S0308814610007235

[75] Lermusieau G, Bulens M, Collin S. Use of GC-olfactometry to identify the hop aromatic compounds in beer. Journal of Agricultural and Food Chemistry(Internet). 2001;**49**(8):3867-3874. Available from: http://dx.doi.org/10.1021/jf0101509

[76] Capozzi V, Makhoul S, Aprea E, Romano A, Cappellin L, Sanchez Jimena A, et al. PTR-MS characterization of VOCs associated with commercial aromatic bakery yeasts of wine and beer origin. Molecules. 2016;**21**(4)

[77] De Clippeleer J, Van Opstaele F, Vercammen J, Francis GJ, De Cooman L, Aerts G. Real-time profiling of volatile malt aldehydes using selected ion flow tube mass spectrometry. LC GC North America [Internet]. 2016:67-74. Available from: http://cat.inist.fr/?aModele=afficheN&cpsidt=23239175. [Accessed: March 1, 2017]